21世纪高等学校规划教材 | 电子信息

电气工程专业英语实用教程（第3版）

张强华　陆巧儿　胡莹　编著

清华大学出版社
北京

内 容 简 介

本书的目的在于切实提高读者的专业英语能力。

本书体例上以 Unit 为单位,每一 Unit 由以下几部分组成:课文——这些课文包括基础知识和基本概念;单词、词组及缩略语——给出课文中出现的新词、常用词组及缩略语,读者由此可以积累电气专业的基本词汇;难句讲解——讲解课文中出现的疑难句子,培养读者的阅读理解能力;习题——既有针对课文的练习,也有一些开放性的练习;科技英语翻译知识——帮助读者掌握基本的专业英语翻译技巧;阅读材料——提供最新的设备和工具软件的相关资料,可进一步扩大读者的视野;参考答案——可供读者检查学习效果。

本书可作为高等院校的专业英语教材,高职高专院校也可选用;作为培训班教材和供从业人员自学亦颇得当。

本书封面贴有清华大学出版社防伪标签,无标签者不得销售。
版权所有,侵权必究。举报: 010-62782989, beiqinquan@tup.tsinghua.edu.cn。

图书在版编目(CIP)数据

电气工程专业英语实用教程/张强华,陆巧儿,胡莹编著. —3 版. —北京:清华大学出版社,2017
(2023.8重印)
(21 世纪高等学校规划教材·电子信息)
ISBN 978-7-302-47315-2

Ⅰ. ①电… Ⅱ. ①张… ②陆… ③胡… Ⅲ. ①电工技术－英语－教材 Ⅳ. ①TM

中国版本图书馆 CIP 数据核字(2017)第 124419 号

责任编辑:魏江江　薛　阳
封面设计:傅瑞雪
责任校对:梁　毅
责任印制:杨　艳

出版发行:清华大学出版社
　　　网　　址:http://www.tup.com.cn, http://www.wqbook.com
　　　地　　址:北京清华大学学研大厦 A 座　　邮　编:100084
　　　社 总 机:010-83470000　　邮　购:010-62786544
　　　投稿与读者服务:010-62776969,c-service@tup.tsinghua.edu.cn
　　　质量反馈:010-62772015,zhiliang@tup.tsinghua.edu.cn
　　　课件下载:http://www.tup.com.cn,010-83470236

印 装 者:北京国马印刷厂
经　　销:全国新华书店
开　　本:185mm×260mm　　印　张:17.75　　字　数:431 千字
版　　次:2016 年 1 月第 1 版　　2017 年 8 月第 3 版　　印　次:2023 年 8 月第 10 次印刷
印　　数:59001~61000
定　　价:49.50 元

产品编号:074706-02

出 版 说 明

随着我国改革开放的进一步深化,高等教育也得到了快速发展,各地高校紧密结合地方经济建设发展需要,科学运用市场调节机制,加大了使用信息科学等现代科学技术提升、改造传统学科专业的投入力度,通过教育改革合理调整和配置了教育资源,优化了传统学科专业,积极为地方经济建设输送人才,为我国经济社会的快速、健康和可持续发展以及高等教育自身的改革发展做出了巨大贡献。但是,高等教育质量还需要进一步提高以适应经济社会发展的需要,不少高校的专业设置和结构不尽合理,教师队伍整体素质亟待提高,人才培养模式、教学内容和方法需要进一步转变,学生的实践能力和创新精神亟待加强。

教育部一直十分重视高等教育质量工作。2007年1月,教育部下发了《关于实施高等学校本科教学质量与教学改革工程的意见》,计划实施"高等学校本科教学质量与教学改革工程"(简称"质量工程"),通过专业结构调整、课程教材建设、实践教学改革、教学团队建设等多项内容,进一步深化高等学校教学改革,提高人才培养的能力和水平,更好地满足经济社会发展对高素质人才的需要。在贯彻和落实教育部"质量工程"的过程中,各地高校发挥师资力量强、办学经验丰富、教学资源充裕等优势,对其特色专业及特色课程(群)加以规划、整理和总结,更新教学内容、改革课程体系,建设了一大批内容新、体系新、方法新、手段新的特色课程。在此基础上,经教育部相关教学指导委员会专家的指导和建议,清华大学出版社在多个领域精选各高校的特色课程,分别规划出版系列教材,以配合"质量工程"的实施,满足各高校教学质量和教学改革的需要。

为了深入贯彻落实教育部《关于加强高等学校本科教学工作,提高教学质量的若干意见》精神,紧密配合教育部已经启动的"高等学校教学质量与教学改革工程精品课程建设工作",在有关专家、教授的倡议和有关部门的大力支持下,我们组织并成立了"清华大学出版社教材编审委员会"(以下简称"编委会"),旨在配合教育部制定精品课程教材的出版规划,讨论并实施精品课程教材的编写与出版工作。"编委会"成员皆来自全国各类高等学校教学与科研第一线的骨干教师,其中许多教师为各校相关院、系主管教学的院长或系主任。

按照教育部的要求,"编委会"一致认为,精品课程的建设工作从开始就要坚持高标准、严要求,处于一个比较高的起点上。精品课程教材应该能够反映各高校教学改革与课程建设的需要,要有特色风格、有创新性(新体系、新内容、新手段、新思路,教材的内容体系有较高的科学创新、技术创新和理念创新的含量)、先进性(对原有的学科体系有实质性的改革和发展,顺应并符合21世纪教学发展的规律,代表并引领课程发展的趋势和方向)、示范性(教材所体现的课程体系具有较广泛的辐射性和示范性)和一定的前瞻性。教材由个人申报或各校推荐(通过所在高校的"编委会"成员推荐),经"编委会"认真评审,最后由清华大学出版社审定出版。

目前，针对计算机类和电子信息类相关专业成立了两个"编委会"，即"清华大学出版社计算机教材编审委员会"和"清华大学出版社电子信息教材编审委员会"。推出的特色精品教材包括：

（1）21世纪高等学校规划教材·计算机应用——高等学校各类专业，特别是非计算机专业的计算机应用类教材。

（2）21世纪高等学校规划教材·计算机科学与技术——高等学校计算机相关专业的教材。

（3）21世纪高等学校规划教材·电子信息——高等学校电子信息相关专业的教材。

（4）21世纪高等学校规划教材·软件工程——高等学校软件工程相关专业的教材。

（5）21世纪高等学校规划教材·信息管理与信息系统。

（6）21世纪高等学校规划教材·财经管理与应用。

（7）21世纪高等学校规划教材·电子商务。

（8）21世纪高等学校规划教材·物联网。

清华大学出版社经过三十多年的努力，在教材尤其是计算机和电子信息类专业教材出版方面树立了权威品牌，为我国的高等教育事业做出了重要贡献。清华版教材形成了技术准确、内容严谨的独特风格，这种风格将延续并反映在特色精品教材的建设中。

清华大学出版社教材编审委员会
联系人：魏江江
E-mail：weijj@tup.tsinghua.edu.cn

前言

当今,电气行业的新技术、新设备和新工具不断出现,要掌握这些新知识和新技能,从业人员就必须不断地学习,这有赖于专业英语水平的提高。为此,就必须进行针对性的专门学习。本书的目的就在于切实提高读者实际使用电气专业英语的能力。

本书体例上以 Unit 为单位,每一个 Unit 由以下几部分组成:课文——内容包括了基础知识和基本概念;单词、词组及缩略语——给出课文中出现的新词、常用词组及缩略语,读者由此可以积累电气专业方面的基本词汇;难句讲解——讲解课文中出现的疑难句子,培养读者的阅读理解能力;习题——既有针对课文的练习,也有一些开放性的练习;科技英语翻译知识——帮助读者掌握基本的专业英语翻译技巧;阅读材料——提供最新的设备和工具软件的相关资料,可进一步扩大读者的视野;参考答案——可供读者检查学习效果。

本书适合作为电气类专业的教材,适应当前许多院校已经实行的"宽口径"人才培养模式。

本书新增了目前工程人员最需要的电气 CAD 软件的内容,在结构上也非常适合组织教学,词汇加注了音标。

本书在编写中,着重从"教师教什么"、"学生就业后用什么"来考虑并结合学生的具体情况,针对学生毕业后的就业环境,根据未来工作实际的要求,对内容做了切合实际的精心安排。

在学习本书过程中,如有任何问题,可以通过电子邮件与我们交流。我们一定会给予答复。如果读者没有收到回复,请再次联系。邮件标题请注明姓名及"电气工程专业英语实用教程(清华大学版)"字样,否则会被当作垃圾邮件删除。教师也可索取参考试卷。

我们的 E-mail 地址如下:
zqh3882355@sina.com;zxdningbo@etang.com。

望大家不吝赐教,使本书成为一部"符合学生实际、切合行业实况、知识实用丰富、严谨开放创新"的优秀教材。

本书既可作为高等院校的专业英语教材,也可供高职高专院校选用,还可作为培训班教材和供从业人员自学,亦颇得当。

编　者

第三版前言

本书出版以来,承蒙读者厚爱,一印再印。借此机会,向读者朋友表示衷心感谢!

本次修订,主要进行了以下工作。

(1) 修订了上一版本中的若干错误。

(2) 增加了一套自测题,以便读者自我检查并熟悉考试的内容与格式。

(3) 将全书单词汇总为词汇表,并增加了一些行业内经常使用的词汇,约计单词 700 个、词组 800 个以及缩写词 200 个。这不仅有助于读者学习本课程,也有益于读者在以后工作中长期查阅。

(4) 更新部分课文,以适应行业发展。

张强华

2017.5.5

目 录

Unit 1 ··· 1
  Text A   Basic Electrical Concept ··· 1
    New Words and Phrases ··· 4
    Notes ··· 5
    Exercises ·· 6
  Text B   Introduction to AC ·· 9
    New Words and Phrases ··· 13
    Exercises ·· 14
  科技英语翻译知识　词义的选择 ··· 14
      Reading Material ·· 15
  参考译文　电的基本概念 ·· 18

Unit 2 ··· 20
  Text A   Electrical Resistance and Conductance ···························· 20
    New Words and Phrases ··· 23
    Notes ··· 25
    Exercises ·· 26
  Text B   What Is Capacitor? ·· 28
    New Words and Phrases ··· 31
    Exercises ·· 32
  科技英语翻译知识　翻译的标准 ··· 32
      Reading Material ·· 34
  参考译文　电阻和电导 ··· 37

Unit 3 ··· 40
  Text A   Simple Electrical Circuit ··· 40
    New Words and Phrases ··· 44
    Notes ··· 45
    Exercises ·· 46
  Text B   DC Parallel Circuit ·· 48
    New Words and Phrases ··· 53
    Exercises ·· 53

科技英语翻译知识　词义的引申 ·· 55
　　　Reading Material ·· 56
　参考译文　简单电路 ·· 61

Unit 4 ·· 64

　Text A　Basic Semiconductor Crystal Structure ·· 64
　　　New Words and Phrases ·· 66
　　　Notes ·· 67
　　　Exercises ·· 67
　Text B　The PN Junction ·· 70
　　　New Words and Phrases ·· 71
　　　Exercises ·· 72
　科技英语翻译知识　词义的增减 ·· 73
　　　Reading Material ·· 75
　参考译文　基本半导体晶体结构 ·· 81

Unit 5 ·· 83

　Text A　Number Systems ·· 83
　　　New Words and Phrases ·· 87
　　　Notes ·· 87
　　　Exercises ·· 88
　Text B　Digital Circuit Elements ·· 90
　　　New Words and Phrases ·· 94
　　　Exercises ·· 95
　科技英语翻译知识　词类的转换 ·· 95
　　　Reading Material ·· 97
　参考译文　数字系统 ·· 104

Unit 6 ·· 106

　Text A　AC Motors ·· 106
　　　New Words and Phrases ·· 110
　　　Notes ·· 112
　　　Exercises ·· 113
　Text B　Basic DC Motor Operation ·· 116
　　　New Words and Phrases ·· 119
　　　Exercises ·· 120
　科技英语翻译知识　否定的译法 ·· 120
　　　Reading Material ·· 122
　参考译文　交流电动机 ·· 125

Unit 7 128

Text A The Basis of Control 128
   New Words and Phrases 131
   Notes 132
   Exercises 133

Text B Digital Control Systems 135
   New Words and Phrases 138
   Exercises 140

科技英语翻译知识 被动语态的译法 140
   Reading Material 142

参考译文 控制基础 146

Unit 8 148

Text A PLC 148
   New Words and Phrases 153
   Notes 154
   Exercises 154

Text B SPICE 157
   New Words and Phrases 160
   Exercises 162

科技英语翻译知识 从句的译法 162
   Reading Material 165

参考译文 可编程逻辑控制器 173

Unit 9 176

Text A What Is CNC? 176
   New Words and Phrases 178
   Notes 179
   Exercises 180

Text B The Basics of Computer Numerical Control 183
   New Words and Phrases 190
   Exercises 191

科技英语翻译知识 汉语四字格的运用 191
   Reading Material 193

参考译文 CNC是什么? 197

Unit 10 200

Text A Industrial Bus 200

　　　　New Words and Phrases ··· 203
　　　　Notes ··· 204
　　　　Exercises ·· 205
　　Text B　Serial Communications Systems ·· 208
　　　　New Words and Phrases ··· 210
　　　　Exercises ·· 211
　　科技英语翻译知识　篇章翻译 ··· 212
　　　　Reading Material ·· 213
　　参考译文　工业总线 ··· 218

附录1　自测题 ··· 221

附录2　词汇总表 ··· 229

Unit 1

Text A Basic Electrical Concept

1. Electric Charges

(1) Neutral state of an atom

Elements are often identified by the number of electrons in orbit around the nucleus of the atoms that making up the element and by the number of protons in the nucleus. [1] A hydrogen atom, for example, has only one electron and one proton. An aluminum atom (illustrated) has 13 electrons and 13 protons. An atom with an equal number of electrons and protons is said to be electrically neutral.

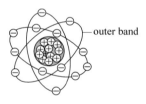

(2) Positive and negative charges

Electrons in the outer band of an atom are easily displaced by the application of some external force. Electrons which are forced out of their orbits can result in a lack of electrons where they leave and an excess of electrons where they come to rest. [2] The lack of electrons is called a positive charge because there are more protons than electrons. The excess of electrons has a negative charge. Positive or negative charge is caused by absence or excess of electrons. The number of protons remains constant.

Neutral charge   Negative charge   Positive charge

(3) Attraction and repulsion of electric charges

The old saying, "opposites attract" is true when dealing with electric charges. Charged bodies have an invisible electric field around them. When two like-charged bodies are brought together, their electric field will work to repel them. When two unlike-charged bodies are brought together, their electric field will work to attract them. The electric field around a charged body is represented by invisible lines of force. The invisible lines of force represent an invisible electrical field that causes the attraction and repulsion. Lines of force are shown leaving a body with a positive charge and entering a body with a negative charge.

2. Current

Electricity is the flow of free electrons in a conductor from one atom to the next atom in the same general direction. This flow of electrons is referred to as current and is designated by the symbol "I". Electrons move through a conductor at different rates so electric current has different values. Current is determined by the number of electrons that pass through a cross-section of a conductor in one second. We must remember that atoms are very small. It takes about 1 000 000 000 000 000 000 000 000 atoms to fill one cubic centimeter of a copper conductor. This number can be simplified using mathematical exponents. Instead of writing 24 zeros after the number 1, write 10^{24}. Trying to measure even small values of current would result in unimaginably large numbers.[3] For this reason current is measured in Amperes which is abbreviated "Amp". The symbol for amp is the letter "A". A current of one amp means that in one second about 6.24×10^{18} electrons move through a cross-section of conductor. These numbers are given for information only and you do not need to be concerned with them. It is important, however, that the concept of current flow be understood.

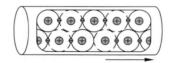

(1) Units of measurement for current

The following chart reflects special prefixes that are used when dealing with very small or large values of current:

Prefix	Symbol	Decimal
1 kiloampere	1kA	1 000A
1 milliampere	1mA	1/1 000A
1 microampere	1μA	1/1 000 000A

(2) Direction of current flow

Some authorities distinguish between electron flow and current flow. Conventional current flow theory ignores the flow of electrons and states that current flows from positive to negative. To avoid confusion, We will use the electron flow concept which states that electrons flow from negative to positive.

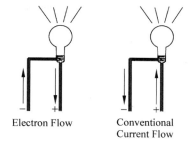

Electron Flow   Conventional Current Flow

3. Voltage

Electricity can be compared with water flowing through a pipe. A force is required to get water to flow through a pipe. This force comes from either a water pump or gravity. Voltage is the force that is applied to a conductor that causes electric current to flow. Electrons are negative and are attracted by positive charges. They will always be attracted from a source having an excess of electrons, thus having a negative charge, to a source having a deficiency of electrons which has a positive charge.[4] The force required to make electricity flow through a conductor is called a difference in potential, electromotive force(emf), or more simply referred to as voltage. Voltage is designated by the letter "U", or the letter "V". The unit of measurement for voltage is volt which is designated by the letter "V".

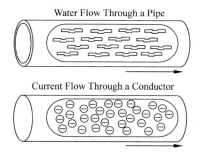

Water Flow Through a Pipe

Current Flow Through a Conductor

(1) Voltage sources

An electric voltage can be generated in various ways. A battery uses electrochemical process to generate voltage. A car's alternator and a power plant generator utilize a magnetic induction process. All voltage sources share the characteristic of an excess of electrons at one terminal and a shortage at the other terminal. This results in a difference of potential between the two terminals.

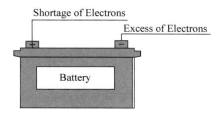

(2) Voltage symbol

The terminals of a battery are indicated symbolically by an electrical drawing of two lines. The longer line indicates the positive terminal. The shorter line indicates the negative.

(3) Units of measurement for voltage

The following chart reflects special prefixes that are used when dealing with very small or large values of voltage:

Prefix	Symbol	Decimal
1 kilovolt	1kV	1 000V
1 millivolt	1mV	1/1 000V
1 microvolt	1μV	1/1 000 000V

New Words and Phrases

atom	['ætəm]	n.	原子
neutral state			中性状态
orbit	['ɔːbit]	n.	轨道；轨迹
designate	['dezigneit]	vt.	指定(出示),标明
nucleus	['njuːkliəs]	n.	核子
cubic	['kjuːbik]	adj.	立方体的,立方的
proton	['prəutɔn]	n.	质子
force out			挤(出去),冲(出去)
abbreviate	[ə'briːvieit]	vt.	节略,省略,缩写
excess	[ik'ses]	n.	过度,多余,超过,超额
		adj.	过度的,额外的
conventional	[kən'venʃənl]	adj.	惯例的,常规的,习俗的,传统的
gravity	['græviti]	n.	地心引力,重力
charge	[tʃɑːdʒ]	n.	负荷,电荷,费用,充电
negative charge			负电荷
positive charge			正电荷
absence	['æbsəns]	n.	缺乏,没有
remain	[ri'mein]	vi.	剩余,残存
attraction	[ə'trækʃən]	n.	吸引,吸引力
repulsion	[ri'pʌlʃən]	n.	排斥
electric field			电场
repel	[ri'pel]	vt.	排斥
represent	[ˌriːpri'zent]	vt.	表示,表现
free electron			自由电子
symbol	['simbəl]	n.	符号,记号
rate	[reit]	n.	比率,速度,等级
determine	[di'təːmin]	vt. & vi.	决定,确定,测定

cross-section			截面,断面
exponent	[eks'pəunənt]	n.	指数,幂
unimaginably	[ˌʌni'mædʒinəbl]	adj.	不能想象的,难以理解的
concern with			使关心
current flow			电流
measurement	['meʒəmənt]	n.	量度,测量法
prefix	['priːfiks]	n.	前缀
distinguish	[dis'tiŋgwiʃ]	vt. & vi.	区别,辨别
ignore	[ig'nɔː]	vt.	忽略,不理睬,忽视
electrochemical	[iˌlektrəu'kemikəl]	adj.	电气化学的
alternator	['ɔːltə(ː)neitə]	n.	交流发电机
generator	['dʒenəreitə]	n.	发电机
magnetic	[mæg'netic]	adj.	磁的,有磁性的,有吸引力的
deficiency	[di'fiʃənsi]	n.	缺乏,不足
potential	[pə'tenʃ(ə)l]	adj.	势的,位的
utilize	['juːtilaiz]	vt.	利用
terminal	['təːminl]	n.	终端,接线端,电路接头
characteristic	[ˌkæriktə'ristik]	adj.	特有的,典型的
		n.	特性,特征
shortage	['ʃɔːtidʒ]	n.	不足,缺乏
symbolically	[sim'bɔlikli]	adv.	象征性地
kilovolt	['kiləuvəult]	n.	千伏特
millivolt	[milivəult]	n.	毫伏(特)[=1/1 000 伏(特),略作 mV]
microvolt	[ˌmaikrəu'vəult]	n.	微伏[等于1伏(特)的百万分之一]

Notes

[1] Elements are often identified by the number of electrons in orbit around the nucleus of the atoms making up the element and by the number of protons in the nucleus.

本句中,介词 by 的宾语由两部分组成,一个是 the number of electrons in orbit around the nucleus of the atoms making up the element;另一个是 the number of protons in the nucleus。第一个宾语中,making up the element 修饰 atoms,介词短语 in orbit around the nucleus of the atoms 作定语,修饰 electrons。be identified by 是"通过……鉴别"、"识别"的意思。

元素是通过组成该原子的核子轨道上的电子数及核子中的质子数来区分的。

[2] Electrons which are forced out of their orbits can result in a lack of electrons where they leave and an excess of electrons where they come to rest.

本句中的主语是 electrons,谓语是 can result in, a lack of electrons where they leave and an excess of electrons where they come to rest 是宾语。这个宾语由 and 连接的名词短语 a lack of electrons 以及 an excess of electrons 组成,两个 where 引导地点状语,在理解时要分清这里面的层次。

电子从轨道移走而造成该原子缺少电子,而移入电子的原子会使电子过剩。

[3] Trying to measure even small values of current would result in unimaginably large numbers.

本句的主语是一个动名词短语 trying to measure even small values of current,谓语是动词词组 result in,result in 不能机械地理解为"导致",因为"导致"带有明显的感情色彩,理解为"造成"更顺畅。

试图测量电流微小的值将产生难以想象的巨大的数字。

[4] They will always be attracted from a source having an excess of electrons, thus having a negative charge, to a source having a deficiency of electrons which has a positive charge.

本句的主干是 they will always be attracted from…to…。having an excess of the electrons, thus having a negative charge 两个定语修饰的是第一个 source。having a deficiency of electrons which has a positive charge 修饰的是第二个 source。第二个修饰语虽然没有明显表达出两个定语之间的逻辑关系,但是从第一个修饰语中两个定语用 thus 连接可以判断出,它们仍然有因果关系:因为缺乏电子,才形成正电荷。所以在理解时要注意这层意思。

Exercises

【Ex.1】 根据课文内容,回答以下问题。

(1) What is a hydrogen atom made up of?

(2) Why does an atom present neutral?

(3) How does an atom have positive charge?

(4) What will happen when two like-charged bodies are brought together?

(5) What is defined about the current according to conventional current flow theory?

【Ex. 2】 根据下面的英文解释,写出相应的英文词汇。

英 文 解 释	词 汇
a stable subatomic particle in the lepton family having a rest mass of 9.1066×10^{-28} gram and a unit negative electric charge of approximately 1.602×10^{-19} coulomb	
a colorless, highly flammable gaseous element, the lightest of all gases and the most abundant element in the universe, used in the production of synthetic ammonia and methanol, in petroleum refining, Atomic number 1	
the net measure of this property possessed by a body or contained in a bounded region of space	
a part or particle considered to be an irreducible constituent of a specified system	
having a volume equal to a cube whose edge is of a stated length	
an electric generator that produces alternating current	
of or relating to magnetism or magnets	
the natural force of attraction exerted by a celestial body, such as Earth, upon objects at or near its surface, tending to draw them toward the center of the body	
a unit of potential difference equal to one thousandth of a volt	
a building or group of buildings for the manufacture of a product	

【Ex. 3】 把下列句子翻译为中文。

(1) A glass rod becomes charged when rubbed with silk, as does a hard-rubber rod when rubbed with fur.

(2) People found that charges produce forces of repulsion and attraction. These are usually small forces.

(3) Negative charges repel negative charges, positive charge repel positive charges, and positive and negative charges attract each other. In short, like charges repel and unlike charges attract.

(4) On further study of charges scientists found that all negative charges are integer multiples of a certain very small charge.

(5) Scientist also discovered that positive charges are integer multiples of a very small charge, the same charges are an electron—but positive. The proton has this charge.

(6) Rubbing a glass rod with a silk cloth or a hard-rubber rod with fur does not create charge. In general, charge cannot be created or destroyed, a fact called the Law of Conservation of Charge.

(7) Rubbing the glass rod with a silk cloth removes electrons from the rod and puts them on the cloth. This charge transfer causes a charge unbalance on both the rod and cloth.

(8) Electrons of an atom have orbits at different distances from the nucleus. For some atoms the electrons in the farthest orbits have only weak forces binding them to the atoms.

(9) An atom with an unbalance charge is called an ion. Ions are charged particles that would produce a current if they could move.

(10) Even at normal room temperatures the outer electrons in metals receive enough heat energy to become free, especially for silver, copper, gold, aluminum.

【Ex. 4】 把下列短文翻译成中文。

The charge of an electron or of a proton is much too small to be basic quantity of charge for almost all practical applications. The SI unit of charge is the coulomb, with the symbol C. A coulomb of negative charge equals that of 6.242×10^{18} electrons. The coulomb is a derived SI unit, which means that it can be derived from SI base units.

【Ex. 5】 通过Internet查找资料,借助如"金山词霸"等电子词典和辅助翻译软件,完成以下技术报告。通过E-mail发送给老师,并附上你收集资料的网址。

(1) 当前世界上有哪些流行的工业控制软件,以及它们的主要技术指标(附上各种最新产品的主要界面图片)。

(2) 当前世界上有哪些流行的电路设计软件,以及它们的主要技术指标(附上各种最新产品的主要界面图片)。

Text B Introduction to AC

The supply of current for electrical devices may come from a direct current source (DC), or an alternating current source (AC). In direct current electricity, electrons flow continuously in one direction from the power source through a conductor to a load and back to the power source. The voltage in direct current remains constant. DC power sources include batteries and DC generators. In alternating current an AC generator is used to make electrons flow first in one direction then in another. Another name for an AC generator is an alternator. The AC generator reverses terminal polarity many times a second. Electrons will flow through a conductor from the negative terminal to the positive terminal, first in one direction then another.

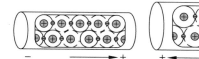

AC

1. AC sine wave

Alternating voltage and current vary continuously. The graphic representation for AC is sine wave. A sine wave can represent current or voltage. There are two axes. The vertical axis represents the direction and magnitude of current or voltage. The horizontal axis represents time.

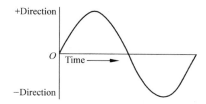

When the waveform is above the time axis, current is flowing in one direction. This is referred to as the positive direction. When the waveform is below the time axis, current is flowing in the opposite direction. This is referred to as the negative direction. A sine wave moves through a complete rotation of 360°, which is referred to as one cycle. Alternating current goes through many of these cycles in one second. The unit of measurement of cycles per second is hertz(Hz). In the United States alternating current is usually generated at 60Hz.

2. Single-phase and three-phase AC power

Alternating current is divided into single-phase and three-phase types. Single-phase power is used for small electrical demands such as found in the home. Three-phase power is used where large amounts of power are required, such as found in commercial applications and industrial plants. Single-phase power is shown in the above illustration. Three-phase power, as shown in the following illustration, is a continuous series of three overlapping AC cycles. Each wave represents a phase, and is offset by 120 degrees.

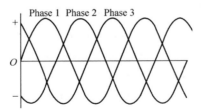

3. Frequency

The number of cycles per second made by voltage induced in the armature is decided by the frequency of the generator. If the armature rotates at a speed of 60 revolutions per second, the generated voltage will be 60 cycles per second. The accepted term for cycles per second is Hertz. The standard frequency in the United States is 60Hz. The following illustration shows 15 cycles in 1/4 second which is equivalent to 60 cycles in one second.

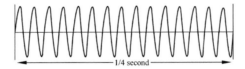

4. Four-pole AC generator

The frequency is the same number of rotations per second if the magnetic field is produced by only two poles. An increase in the number of poles, would cause an increase in the number of cycles completed in a revolution. A two-pole generator would complete one cycle per revolution and a four-pole generator would complete two cycles per revolution. An AC generator produces one cycle per revolution for each pair of poles.

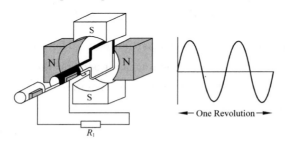

Voltage and Current

1. Peak value

The sine wave illustrates how voltage and current in an AC circuit rise and fall with time.

The peak value of a sine wave occurs twice each cycle, once at the positive maximum value and once at the negative maximum value.

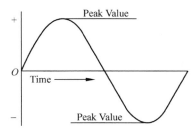

2. Peak-to-peak value

The value of the voltage or current between the peak positive and peak negative values is called the peak-to-peak value.

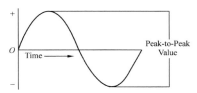

3. Instantaneous value

The instantaneous value is the value at any particular time. It can be in the range of anywhere from zero to the peak value.

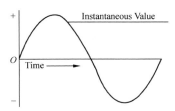

4. Calculating instantaneous voltage

The voltage waveform produced as the armature rotates through 360° is called a sine wave because instantaneous voltage is related to the trigonometric function called sine ($\sin\theta$ = sine of the angle). The sine curve represents a graph of the following equation

$$e = E_{peak} \times \sin\theta$$

Instantaneous voltage is equal to the peak voltage times the sine of the angle of the generator armature. The sine value is obtained from trigonometric tables. The following table reflects a few angles and their sine values.

Angle(θ)/°	$\sin\theta$	Angle(θ)/°	$\sin\theta$
30	0.5	210	-0.5
60	0.866	240	-0.866
90	1	270	-1
120	0.866	300	-0.866
150	0.5	330	-0.5
180	0	360	0

The following example illustrates instantaneous values at 90°, 150°, and 240°. The peak voltage is equal to 100V. By multiplying the sine at the instantaneous angle value, the instantaneous voltage can be calculated.

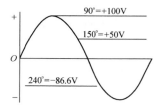

Any instantaneous value can be calculated. For example 240°
$$e = 100 \times (-0.866) = -86.6\text{V}$$

5. Effective value of an AC sine wave

Alternating voltage and current are constantly changing values. A method of translating the varying values into an equivalent constant value is needed. The effective value of voltage and current is the common method of expressing the value of AC. This is also known as the RMS (root-mean-square) value. If the voltage in the average home is said to be 120V, this is the RMS value. The effective value figures out to be 0.707 times the peak value.

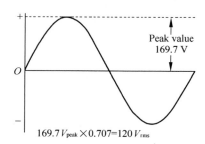

The effective value of AC is defined in terms of an equivalent heating effect when compared to DC. One RMS ampere of current flowing through a resistance will produce heat at the same rate as a DC ampere. For purpose of circuit design, the peak value may also be needed. For example, insulation must be designed to withstand the peak value, not just the effective value. It may be that only the effective value is known. To calculate the peak value, multiply the effective value by 1.41. For example, if the effective value is 100V, the peak value is 141V.

New Words and Phrases

DC (Direct Current)			直流电
reverse	[ri'və:s]	vt.	改变；颠倒，翻转
polarity	[pəu'læriti]	n.	极性
sine	[sain]	n.	正弦
continuously	[kən'tinjuəsli]	adv.	不断地，连续地
graphic representation			图示
sine wave			正弦波
vertical	['və:tikəl]	adv.	垂直的
		n.	垂直线，垂直面，竖向
magnitude	['mægnitju:d]	n.	大小，数量级
horizontal	[,hɔri'zɔntl]	adj.	水平的
waveform	['weivfɔ:m]	n.	波形
referred to as			称为
positive direction			正向
opposite direction			反向
negative direction			负向
cycle	['saɪk(ə)l]	n.	周期
Hertz	['hə:ts]	n.	赫(Hz)，赫兹
phase	[feiz]	n.	相，相位
three-phase			三相
divide into			分为
offset	['ɔ:fset]	n.	偏移量
frequency	['fri:kwənsi]	n.	频率，周率
overlapping	['əuvə'læpiŋ]	vt. & vi.	重叠绕包，搭接
armature	['ɑ:mətjuə]	n.	电枢(电机的部件)
pole	[pəul]	n.	极，磁极，电极
magnetic field			磁场
peak	[pi:k]	n.	顶点
peak value			峰值
maximum	['mæksiməm]	n.	最大量，最大值，极大，极大值
instantaneous	[,instən'teinjəs]	adj.	瞬间的，即刻的，即时的
trigonometric	[,trigənə'metrik]	adj.	三角法的，据三角法的
trigonometric function			三角函数
effective	[i'fektiv]	adj.	有效的
root-mean-square			均方根(值)
withstand	[wið'stænd]	vt.	经受住，抵挡

Exercises

【Ex. 6】 根据文章所提供的信息判断正误。

(1) In direct current electricity, electrons flow continuously in one direction from the source of power through a conductor to a load and back to the source of power.

(2) In alternating current an AC generator is used to make electrons flow continuously in one direction.

(3) Electrons will flow through a conductor from the positive terminal to the negative terminal.

(4) The graphic representation for AC is a sine wave.

(5) In China alternating current is usually generated at 60Hz.

(6) A sine wave moves through a complete rotation of 360°, which is referred to as one cycle.

(7) The peak value of a sine wave occurs once each cycle.

(8) The instantaneous value is the value at any particular time. It can be in the range of anywhere from zero to the peak value.

(9) Instantaneous voltage is equal to the peak voltage times the angle of the generator armature.

(10) For example, if the effective value is 110V, the peak value is 220V.

科技英语翻译知识　词义的选择

在将英语译为中文时,经常遇到确定词义的问题,也就是词义选择的问题。英汉两种语言都有一词多类、一词多义的现象。一词多类就是指一个词往往属于几个词类,具有几个不同的意思。一词多义就是说一个词在同一个词类中,又往往有几个不同的词义。词义的选择可以从以下两个方面着手。

1. 根据词类选择词义

词性不同,所代表的词义往往也不同。当译者遇到某个词时,首先要弄清这个词在句子中的词性,最后才能确定它的词义。例如:

(1) An electron is an extremely small corpuscle with negative charge which **rounds** about the nucleus of an atom.

电子是绕着原子转动带有负电荷的极其微小的粒子。(round 为动词)

(2) The earth goes **round** the sun.

地球环绕太阳运动。(round 为介词)

(3) **Round** surface reflector is a key unit for the solar energy device.

弯曲面反射器是太阳能装置的关键元件。(round 为形容词)

(4) The tree measures about one meter **round**.

这棵树树围约一米。(round 为副词)

(5) This is the whole **round** of knowledge.

这就是全部的知识范围。(round 为名词)

(6) Plastic was at first **based** on coal and wool.

最初塑料是从煤和木材中提取的。(round 为动词)

(7) As we all know, a **base** reacts with an acid to form a salt.

众所周知,碱与酸起反应变成盐。(round 为名词)

(8) Iron and brass are **base** metals.

铁和黄铜为非贵金属。(round 为形容词)

2. 根据上下文选择词义

英语中的同一个单词往往包含几个意思,在不同主题的文章中,表达的意思可能不同。因此在确定某个词的词性后,要根据上下文并结合专业知识来确定它的词义。例如下面例句中的 power:

(1) The electronic microscope possesses very high resolving **power** compared with the optical microscope.

与光学显微镜相比,电子显微镜具有极高的分辨率。

(2) **Power** can be transmitted over a long distance.

电力可以输送到很远的地方。

(3) The fourth **power** of three is eighty-one.

3 的 4 次方是 81。

(4) The combining **power** of one element in the compound must equal the combining power of the other element.

化合物中一种元素的化合价必须等于另一个元素的化合价。

(5) The medical profession enormous **power** to fight disease and sickness has been given by the explosive technological development since 1940.

1940 年以来,随着技术的迅速发展,医学界大大提高了战胜疾病的能力。

(6) Semiconductor devices have no filament or heaters and therefore require no heating **power** or warm up time.

半导体器件没有灯丝和加热器,因此不需要加热功率或时间。

(7) A car needs a lot of **power** to go fast.

汽车高速行驶需要很大的动力。

(8) Stream and waterfall are suitable for the development of hydroelectric **power**.

溪流和瀑布适合用于开发水电能源。

Reading Material

阅读下列文章。

Text	Note
OrCAD New Function OrCAD Capture® design entry is the most widely used schematic entry system in electronic design today for one simple reason: Fast and universal[1] design entry. From designing a new analog circuit, or revising[2] schematic diagrams for an existing PCB, to hierarchical[3] design block—the latest version of OrCAD Capture provides everything you need to expedite[4] your design from verification to manufacturing. OrCAD Capture is now available with an optional component information system (CIS). **1. What's New in Version 10.0?** (1) Unlimited undo/redo Schematic Editor Unlimited[5] undo/redo allows users to remove any of their prior commands or design actions even after their schematic design has been saved. A Label State command has also been added giving you flexibility[6] to try "what if?" scenarios to establish pseudo[7] undo bookmarks and automatic undo sequencing. (2) Synchronize hierarchical design blocks and underlying schematics You can now dynamically propagate[8] port changes from a lower level design to an upper level hierarchical block, or update hierarchical block pin changes to match lower level schematic design with a single mouse click. (3) Substantial improvements in annotation and auto-referencing During annotation[9] of a schematic[10] design, Capture CIS will now utilize power connections and unassigned slot references within a package into consideration. Users also have an option to perform incremental annotation based on the sheet number in title block or the sequential ordering of the schematic within project manager. When set up to use auto-referencing mode, Capture CIS will combine symmetrical[11] parts into a single package. In addition, the reference designators will be automatically reassigned during copy and paste of a design section. (4) File locking Design entry management and integrity is always a key element in team design environments. Various team members may open a schematic design simultaneously[12] where only a single person is identified as the owner. Capture CIS ensures that only one user is identified as the owner allowed for read-write mode while others will be allowed to only view the design in read-only mode. (5) OrCAD Layout 10.0 design reuse support Capture CIS supports the latest OrCAD Layout® place and route editor design reuse function. The design reuse support allows PCB designers using Capture to reuse placement[13] and routing information from one set of components on other identical sets of components in OrCAD Layout. **2. Version 10.0 also includes the following 9.2.3 Web Updates** (1) Auto increment reference designator for pasted parts Allows auto-increment referencing when you copy or paste a part in the Schematic Editor. When a user copies a part that has a reference designator assigned	[1] *adj.* 通用的, 普遍的, 全体的 [2] *vt.* 校阅, 校订 [3] *adj.* 分等级的 [4] *vt. & vi.* 加速 [5] *adj.* 无限的, 无约束的 [6] *n.* 弹性, 适应性, 机动性 [7] *adj.* 假的, 冒充的 [8] *vt. & vi.* 繁殖, 传播 [9] *n.* 注解, 评注 [10] *adj.* 示意性的, 纲要的, 概要的 [11] *adj.* 对称的, 均匀的 [12] *adv.* 同时地 [13] *n.* 放置, 布置

to it, and then pastes that part onto the schematic page, OrCAD Capture assigns the next available reference designator value to the pasted part.

(2) Enhancements to OrCAD Capture-Allegro design flow

Users can back annotate the pin swapping[14] information among heterogeneous parts from Allegro® PCB place and route editor—addressing the GAT0028 restriction[15].

[14] *n.* 交换

[15] *n.* 限制,约束

(3) Part filter

A part selection filter allows users to restrict part searches in part library based on specific criteria. For example, PSpice® simulation users can restrict a part library search so that only parts with associated PSpice simulation model will be listed.

(4) Grid spacing

OrCAD Capture uses the pin-to-pin spacing as the grid[16] spacing for the Schematic Page Editor. This allows user to specify the grid spacing as a fraction of the pin-to-pin spacing.

[16] *n.* 格子,栅格

3. Top Benefits of Version 10.0

(1) Convenient Windows interface for fast, simple editing and sharing

OrCAD Capture® design entry offers an easy-to-use clipboard[17] interface to share schematic graphical data with other application programs using the cut, copy, and paste features. You can also open multiple screens in Capture for editing between schematic pages, or copying and pasting designs between projects.

[17] *n.* 剪贴板

(2) Integration with other Cadence products for one seamless solution

OrCAD Capture is the front-end design tool of choice for PSpice® simulator, OrCAD Layout® place and route editor, Allegro® PCB place and route editor, and Cadence® SPECCTRAQuest™ Signal Integrity analysis. It offers tight integration between tools such as cross-highlighting of components and cross-probing[18] of nets. You can also do forward and back annotation with Capture and OrCAD Layout or Capture and Allegro.

[18] *n.* 探通术

(3) Compatible[19] with popular PCB applications

Provides over 30 PCB netlist formats including OrCAD Layout, Allegro, PADS P2K, and Mentor Graphics Board Station without extra cost.

[19] *adj.* 兼容的,一致的

(4) Capture schematics targeted toward analog or digital simulation, FPGA designs, or PCB layout without leaving OrCAD Capture

Integrates with PSpice, Synplicity Synplify, NC VHDL Desktop, NC Verilog® Desktop, OrCAD Layout, Allegro, and SPECCTRAQuest Signal Integrity solutions.

(5) PLD[20] schematic libraries for PLD designs offer a variety of options

OrCAD Capture provides schematic libraries from six FPGA or PLD vendors[21]: Xilinx, Altera, Actel, Lattice, Lucent, and Atmel. This allows you to stay with one schematic entry tool for six different component vendors' applications.

[20] 可编程逻辑电路

[21] *n.* 卖主,供货商

(6) Integration with Synplify and NC[22] Desktop simulators provide limitless options in PCB designs

OrCAD Capture integrates with Synplify and NC VHDL Desktop providing all capabilities necessary for designing, synthesizing[23], and simulating VHDL design projects. The Design Project Manager provides complete control of your FPGA design

[22] 数控

[23] *n.* 综合,合成

with a simple click of the mouse. Upon finishing your FPGA design, the symbol can be generated for PCB schematic to continue on your board design. You also can simulate[24] your board design with NC VHDL Desktop or NC Verilog Desktop.	[24] *vt.* 模拟，模仿

参考译文　电的基本概念

1. 电荷

（1）中性的原子

元素是通过组成该原子的原子核轨道上的电子数及原子核中的质子数来区分的。例如，一个氢原子只有一个电子和一个质子。一个铝原子有13个电子和13个质子。一个有相同数目的电子和质子数的原子电性呈中性。

（2）正负电荷

原子外圈的电子很容易由于受到外力的作用而被移走。电子从轨道移走会造成该原子缺少电子，而移入电子的原子会使电子过剩。由于质子数多于电子数，缺少电子的原子带正电荷。电子过剩的原子带负电荷。电子不足或过剩产生正或负电荷。质子数始终是恒定不变的。

（3）电荷之间的吸引和排斥

"异性相吸"在处理电荷时是正确的。每个电荷四周都有一个看不见的电场。当两个电性相同的电荷靠近时电场将使两电荷相斥，当两个电性不同的电荷靠近时电场将使两电荷相吸。电荷周围的电场用不可见的电力线表示，不可见的电力线表示引起吸引和排斥的不可见的电场。正电荷用离开的电力线表示，负电荷用进入的电力线表示。

2. 电流

电是在导体中从一个原子以相同方向流向下一个原子的自由电子流。这种电子流就是电流，由符号"I"表示。电子以不同的速度流过导体，电流有不同的值。电流的大小由在1s内流过导体横截面的电子数量决定。我们必须记住原子是非常小的。在$1cm^3$的铜导体中约有1 000 000 000 000 000 000 000 000个原子。这个数字可以用数学的指数形式简化，不用写1后的24个0，而写成10^{24}。试图测量如此微小的电流将产生难以想象的巨大数字。所以，电流用安培数来计量，安培的英文可以简写为"Amp"。安培可以用符号"A"表示。1A的电流意味着在1s之内有$6.24×10^{18}$个电子流过导体的截面。这些数字只是信息，并不需要去关心。但是重要的是理解电流的概念。

（1）测量电流的单位

下面的表反映了处理很小和很大的电流时计量单位的词头。

词　头	符　号	十　进　制
1 千安	1kA	1 000A
1 毫安	1mA	1/1 000A
1 微安	1μA	1/1 000 000A

(2) 电流的方向

一些权威将电子流和电流区分,传统上电流理论忽略电子流,而认为电流是从正极流向负极。为了避免混淆,我们使用的电子流概念是指电子从负极流向正极。

3. 电压

电流可与流过管子的水相比较。水流过管子需要一个力的作用,这个力来自水泵或水的重力。电压就是作用于导体上的引起电流流动的力。电子的电性是负的并能被正电荷吸引,电子总是从由于多余电子而呈负电荷的地方被吸引到由于缺乏电子而呈现电荷为正的地方。这种使得电子流过导体的力叫做电势差、电动势或简单地称为电压。电压用字符"U"或"V"表示。量度电压的单位是伏特,伏特用字符"V"表示。

(1) 电压源

可以有多种产生电压的方式。电池利用电化学过程,而汽车的发电机和电厂的发电机是利用电磁感应作用。所有的电压源共有的特性是在一端电子多余并在另一端电子不足,这就导致了在两端具有不同的电动势。

(2) 电压符号

在电路图中画两条线来象征性代表电池的两端。长的线代表电池的正极,短的线代表电池的负极。

(3) 电压量度单位

下面的表反映了处理很小和很大的电压时计量单位的词头。

词 头	符 号	十 进 制
1 千伏	1kA	1 000V
1 毫伏	1mA	1/1 000V
1 微伏	1μA	1/1 000 000V

Unit 2

Text A Electrical Resistance and Conductance

The electrical resistance of an electrical conductor is a measure of the difficulty to pass an electric current through that conductor. The inverse quantity is electrical conductance, and is the ease with which an electric current passes. Electrical resistance shares some conceptual parallels with the notion of mechanical friction. The SI unit of electrical resistance is the Ohm (Ω), while electrical conductance is measured in Siemens (S).

An object of uniform cross section has a resistance proportional to its resistivity and length and inversely proportional to its cross-sectional area. All materials show some resistance, except for superconductors, which have a resistance of zero[1].

The resistance (R) of an object is defined as the ratio of voltage across it (V) to current through it (I), while the conductance (G) is the inverse:

$R = V/I$ $G = I/V = 1/R$

For a wide variety of materials and conditions, V and I are directly proportional to each other, and therefore R and G are constant (although they can depend on other factors like temperature). This proportionality is called Ohm's law, and materials that satisfy it are called ohmic materials.

In other cases, such as a diode or battery, V and I are not directly proportional. The ratio V/I is sometimes still useful, and is referred to as a "static resistance", since it corresponds to the inverse slope of a chord between the origin and an I-V curve. In other situations, the derivative dV/dI may be most useful; this is called the "differential resistance".

1. Introduction

In the hydraulic analogy, current flowing through a wire (or resistor) is like water flowing through a pipe, and the voltage drop across the wire is like the pressure drop that pushes water through the pipe[2]. Conductance is proportional to how much flow occurs for a given pressure, and resistance is proportional to how much pressure is required to achieve a given flow. (Conductance and resistance are reciprocals.)

The voltage drop (i.e., difference between voltages on one side of the resistor and the

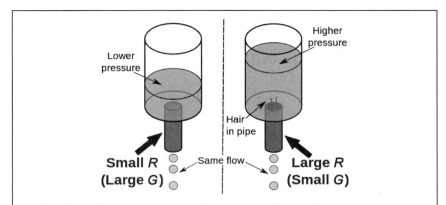

The hydraulic analogy compares electric current flowing through circuits to water flowing through pipes. When a pipe (left) is filled with hair (right), it takes a larger pressure to achieve the same flow of water. Pushing electric current through a large resistance is like pushing water through a pipe clogged with hair: It requires a larger push (electromotive force) to drive the same flow (electric current).

other), not the voltage itself, provides the driving force pushing current through a resistor. In hydraulics, it is similar: The pressure difference between two sides of a pipe, not the pressure itself, determines the flow through it. For example, there may be a large water pressure above the pipe, which tries to push water down through the pipe. But there may be an equally large water pressure below the pipe, which tries to push water back up through the pipe. If these pressures are equal, no water flows.

The resistance and conductance of a wire, resistor, or other element is mostly determined by two properties:

- geometry (shape), and
- material

Geometry is important because it is more difficult to push water through a long, narrow pipe than a wide, short pipe. In the same way, a long, thin copper wire has higher resistance (lower conductance) than a short, thick copper wire.

Materials are important as well. A pipe filled with hair restricts the flow of water more than a clean pipe of the same shape and size. Similarly, electrons can flow freely and easily through a copper wire, but cannot flow as easily through a steel wire of the same shape and size, and they essentially cannot flow at all through an insulator like rubber, regardless of its shape. The difference between, copper, steel, and rubber is related to their microscopic structure and electron configuration, and is quantified by a property called resistivity.

2. Conductors and resistors

Substances in which electricity can flow are called conductors. A piece of conducting material of a particular resistance meant for use in a circuit is called a resistor. Conductors are

A 6.5 MΩ resistor, as identified by its electronic color code (blue–green–black-yellow). An ohmmeter could be used to verify this value.

made of high-conductivity materials such as metals, in particular copper and aluminium. Resistors, on the other hand, are made of a wide variety of materials depending on factors such as the desired resistance, amount of energy that it needs to dissipate, precision, and costs[3].

3. Relation to resistivity and conductivity

The resistance of a given object depends primarily on two factors: What material it is made of, and its shape. For a given material, the resistance is inversely proportional to the cross-sectional area; for example, a thick copper wire has lower resistance than an otherwise-identical thin copper wire. Also, for a given material, the resistance is proportional to the length; for example, a long copper wire has higher resistance than an otherwise-identical short copper wire. The resistance R and conductance G of a conductor of uniform cross section, therefore, can be computed as

$R = \rho(L/A)$

$G = \sigma(A/L)$

where L is the length of the conductor, measured in metres [m], A is the cross-sectional area of the conductor measured in square metres [m^2], σ (sigma) is the electrical conductivity measured in siemens per meter (S·m^{-1}), and ρ is the electrical resistivity (also called specific electrical resistance) of the material, measured in ohm-metres (Ω·m). The resistivity and conductivity are proportionality constants, and therefore depend only on the material the wire is made of, not the geometry of the wire. Resistivity and conductivity are reciprocals: $\rho = 1/\sigma$. Resistivity is a measure of the material's ability to oppose electric current.

This formula is not exact, as it assumes the current density is totally uniform in the conductor, which is not always true in practical situations. However, this formula still provides a good approximation for long thin conductors such as wires.

Another situation for which this formula is not exact is with alternating current (AC), because the skin effect inhibits current flow near the center of the conductor. For this reason, the geometrical cross-section is different from the effective cross-section in which current actually flows, so resistance is higher than expected. Similarly, if two conductors near each other carry AC current, their resistances increase due to the proximity effect. At commercial power frequency, these effects are significant for large conductors carrying large currents, such as busbars in an electrical substation, or large power cables carrying more than a few hundred amperes[4].

4. Measuring resistance

An instrument for measuring resistance is called an ohmmeter. Simple ohmmeters cannot measure low resistances accurately because the resistance of their measuring leads causes a voltage drop that interferes with the measurement, so more accurate devices use four-terminal sensing.

New Words and Phrases

conductance	[kən'dʌktəns]	n.	电导,导体,电导系数
electrical	[i'lektrik(ə)l]	adj.	电的,有关电的
conductor	[kən'dʌktə]	n.	导体
electric current			电流
IS (International System)			国际系统
Ohm	[əum]	n.	欧姆
Siemens	['si:mənz]	n.	西门子(欧姆的倒数)
uniform cross section			等截面
resistivity	[ˌri:zis'tiviti]	n.	电阻系数
proportional	[prə'pɔ:ʃnl]	adj.	比例的,成比例的,相称的
cross-sectional area			断面面积,横截面积
superconductor	[ˌsju:pəkən'dʌktə]	n.	超导(电)体
be defined as			被定义为
voltage	['vəultidʒ]	n.	电压,伏特数
a variety of			多种的
constant	['kɔnstənt]	n.	常数,恒量
		adj.	不变的
temperature	['tempritʃə]	n.	温度
proportionality	[prəˌpɔ:ʃə'næliti]	n.	比例(性)
Ohm's law			欧姆定律
ohmic	['əumik]	adj.	欧姆的
diode	['daiəud]	n.	二极管
battery	['bætəri]	n.	电池
static resistance			静态电阻
slope	[sləup]	n.	斜坡,斜面,倾斜
chord	[kɔ:d]	n.	弦
curve	[kə:v]	n.	曲线,弯曲
		vt.	弯,使弯曲
		vi.	成曲形
differential resistance			微分电阻,内阻
analogy	[ə'nælədʒi]	n.	模拟,类推
electromotive force			电动势

pressure drop		压力降
reciprocal	[ri'siprəkəl]	adj. 倒数的
		n. 倒数
resistor	[ri'zistə]	n. 电阻器
driving force		驱动力
hydraulics	['hai'drɔ:liks]	n. 水力学
pressure difference		压力差,差压
geometry	[dʒi'ɔmitri]	n. 几何形状
shape	[ʃeip]	n. 外形,形状
narrow	['nærəu]	n. 狭窄部分,隘路
		adj. 狭窄的
copper wire		铜线
thick	[θik]	adj. 厚的,粗的
fill with		使充满
electron	[i'lektrɔn]	n. 电子
steel wire		钢丝
insulator	['insjuleitə]	n. 绝缘体,绝热器
rubber	['rʌbə]	n. 橡皮,橡胶
regardless of		不管,不顾
microscopic	[ˌmaikrə'skɔpik]	adj. 用显微镜可见的,精微的
electron configuration		电子构型,电子组态
ohmmeter	['əumˌmi:tə]	n. 欧姆计,电阻表
verify	['verifai]	vt. 检验,校验,查证,核实
substance	['sʌbstəns]	n. 物质
conductivity	[ˌkɔndʌk'tiviti]	n. 传导性,传导率
aluminium	[ˌælju:'minjəm]	n. 铝
		adj. 铝的
dissipate	['disipeit]	v. 消耗
precision	[pri'siʒən]	n. 精确,精密度,精度
make of		用……制造,形成,构成
thin	[θin]	adj. 薄的,细的
identical	[ai'dentikəl]	adj. 同一的,同样的
compute	[kəm'pju:t]	v. 计算,估计
square metres		平方米
sigma	['sigmə]	n. 西格玛,希腊字母(Σ,σ)
electrical conductivity		电导率
proportionality constant		比例常数
oppose	[ə'pəuz]	v. 抵制,反对
exact	[ig'zækt]	adj. 精确的,准确的

current density			电流密度
practical situation			实际情况
formula	[ˈfɔːmjulə]	n.	公式,规则
skin effect			趋肤效应
inhibit	[inˈhibit]	v.	抑制,约束
proximity effect			邻近效应
commercial power frequency			市电频率
busbar	[ˈbʌsbɑː]	n.	汇流排,汇流条
electrical substation			变压站
power cable			电力电缆
ampere	[ˈæmpiə]	n.	安培
instrument	[ˈinstrumənt]	n.	工具,器械,器具
interfere with			妨碍,干涉,干扰
four-terminal sensing			四端子检测,四线检测,四点探针法

Notes

[1] All materials show some resistance, except for superconductors, which have a resistance of zero.

本句中,which have a resistance of zero 是一个非限定性定语从句,对 superconductors 进行补充说明。except for 的意思是"除……之外"。

[2] In the hydraulic analogy, current flowing through a wire (or resistor) is like water flowing through a pipe, and the voltage drop across the wire is like the pressure drop that pushes water through the pipe.

本句中,flowing through a wire (or resistor) 是一个现在分词短语,作定语,修饰和限定 current;flowing through a pipe 也是一个现在分词短语,作定语,修饰和限定 water。that pushes water through the pipe 是一个定语从句,修饰和限定 the pressure drop。

[3] Resistors, on the other hand, are made of a wide variety of materials depending on factors such as the desired resistance, amount of energy that it needs to dissipate, precision, and costs.

本句中,on the other hand 的意思是"另一方面";be made of 的意思是"由……构成";a wide variety of 的意思是"种种,多种多样"。that it needs to dissipate 是一个定语从句,修饰和限定 amount of energy。such as the desired resistance, amount of energy that it needs to dissipate, precision, and costs 对 factors 进行举例说明。

[4] At commercial power frequency, these effects are significant for large conductors carrying large currents, such as busbars in an electrical substation, or large power cables carrying more than a few hundred amperes.

本句中, carrying large currents 是一个现在分词短语,作定语,修饰和限定 large conductors, such as busbars in an electrical substation 是对 large conductors 的举例说明。carrying more than a few hundred amperes 也是一个现在分词短语,作定语,修饰和限定 large power cables。

Exercises

【Ex. 1】 根据课文内容,回答以下问题。

(1) What is the electrical resistance of an electrical conductor?

(2) What provides the driving force pushing current through a resistor? What does the pressure difference between two sides of a pipe determine?

(3) What determines the resistance and conductance of a wire, resistor, or other element?

(4) What are conductors made of?

(5) Why cannot simple ohmmeters measure low resistances accurately?

【Ex. 2】 根据下面的英文解释,写出相应的英文词汇。

英 文 解 释	词 汇
that is the degree to which a substance prevents the flow of an electric current through it	
that is a material or covering which electricity, heat or sound cannot go through	
that allow electricity or heat to go through	
the standard unit of electrical resistance	
the standard unit used to measure how strongly an electrical current is sent around an electrical system	
two pieces of wire were connected end to end	
two pieces of wire were connected side by side	
the standard unit of measurement for the strength of an electrical current	
a standard or accepted way of doing or making something, the items needed for it, or a mathematical rule expressed in a set of numbers and letters	
the standard measure of electrical power	

【**Ex. 3**】 把下列句子翻译为中文。

（1）Resistors are used to control voltages and currents.

（2）Resistors are components that have a predetermined resistance. Resistance determines how much current will flow through a component.

（3）A very high resistance resistor allows very little current to flow.

（4）Sparks and lights are brief displays of current flow through air. The light is created as the current burns parts of the air.

（5）Metals have very low resistance. A low resistance allows a large amount of current to flow.

（6）That is why wires are made of metal. They allow current to flow from one point to another point without any resistance. Wires are usually covered with rubber or plastic.

（7）High voltage power lines are covered with thick layers of plastic to make them safe, but they become very dangerous when the line breaks and the wire is exposed and no longer separated from other things by insulation.

（8）Making the resistance higher will let less current flow so the volume goes down.

（9）Variable resistors are also common components. They have a dial or a knob that allows you to change the resistance.

（10）For example, a 500Ω variable resistor can have a resistance of anywhere between 0Ω and 500Ω.

【Ex. 4】 把下列短文翻译成中文。

Resistance is given in units of ohms(Ω). (Ohms are named after Mho Ohms who played with electricity as a young boy in Germany.) Common resistor values are from 100Ω to 100 000Ω. Each resistor is marked with colored stripes to indicates its resistance. To learn how to calculate the value of a resistor by looking at the stripes on the resistor.

【Ex. 5】 通过 Internet 查找资料，借助如"金山词霸"等电子词典和辅助翻译软件，完成以下技术报告。通过 E-mail 发送给老师，并附上收集资料的网址。

（1）当前世界上有哪些最主要的电阻器件生产厂家以及有哪些最新型的产品（附各种最新产品的图片）。

（2）当前世界上有哪些最主要的电容器件生产厂家以及有哪些最新型的产品（附各种最新产品的图片）。

Text B What Is Capacitor?

The letter C stands for capacitance. A capacitor is formed by two or more conductive parallel plates separated by an insulating material or dielectric. Typical dielectrics you will encounter are air, mica, ceramic or plastic.

Consider an air variable capacitor used either as a trimmer or even a tuning element in a radio. It is merely two or more conductive plates separated by air. —Is this not so?

Go back to our inductors above, could a similar description applied to them?

That is why inductors also have capacitance! We have also learnt a small piece of wire has inductance. This is why capacitors exhibit inductance.

At some point our inductor with its inherent capacitance (called stray) will resonate (you will learn about that later) and this is called self resonant frequency. The same rule applies to capacitors. Mostly it is only the VHF and UHF aficionado who have to greatly concern about these properties. Stray capacitance is everywhere. Sometimes it can be used as advantage, usually you take it into account (that is another function of trimmers) but often it's a monumental pain.

I said mostly—which means if you forget about it, then I can surely guarantee one day this property will bite you. That will be the day when you're working on your pet project, which of course won't work as expected and you don't know why your theory doesn't work out in the real world—another "gotcha".

Stray capacitance in sloppy layouts account for unexpected oscillations, no oscillations, different circuit responses and generally cause you to "become a victim of the bottle, eternally caught in the grip of the grape and, if you become like me, occasionally ruin an otherwise good keyboard with an even much better claret."

Even the best of designs and careful layouts are affected. Not only stray capacitance but also stray inductance can affect you. Try some high speed, double sided, digital PCB designs.

Back to capacitors. Apart from their uses in resonant circuits they are used as:

(1) **DC blocking devices**—A capacitor will pass AC or RF but not DC. That is often the function of coupling capacitors in circuits. The coupling capacitor will pass our required signal and block the DC supply from the previous stage e.g. $0.001\mu F$ ($1\,000pF$ or $1nF$).

(2) **supply by-pass capacitor**—A capacitor used to pass AC or RF. Examples including by-pass of a DC supply to an emitter by-pass capacitor. When used on a DC supply line the capacitor shunts (shorts) to ground any unwanted AC or RF to avoid contamination of our supply e.g. $0.1\mu F$ ($100\,000pF$ or $100nF$).

(3) **reservoir by-pass capacitor**—Used in the output of a DC rectifier to smooth out the power line AC pulses and as a reservoir between the charging pulses. Think of it as a big water storage tank. This is where you have the high value capacitors e.g. $10\,000\mu F/63V$ ($10\,000U$).

(4) **emitter by-pass capacitor**—In the case of an emitter by-pass our transistor may, as one example, be considered as two distinct models, one under DC conditions which sets up how we want the transistor to operate and another under AC or RF conditions e.g. amplifier. In this case our emitter by-pass capacitor merely makes any emitter resistor invisible for AC or RF purposes, typical values might range from $2.2\mu F$ down to $0.1\mu F$. (2U2 down to 100N)

Parallel plate capacitors have a formula for calculating capacitance
$$C = 0.22248 \times K \times (A/t)$$
where

C = capacitance in pF.

K = dielectric constant (air = 1).

A = area of plates or dielectric.

t = thickness of dielectric or spacing in the case of air variables.

(5) **series or parallel capacitors**—Series and parallel combinations calculate in an opposite fashion to inductors, i.e. parallel are
$$C_{total} = C_1 + C_2 + \cdots + C_n$$
and series are
$$C_{total} = \{1/[(1/C_1) + (1/C_2) + (1/C_n)]\}$$

Now here come some very useful formulas using the one above you immediately should determine that a 22pF capacitor in series with a 15pF capacitor gives a total of 8.92pF. Go ahead do it! Check me out.

What if we had say 105pF already in a circuit and needed to reduce it to say 56pF. A real world example might be say the only air variable capacitor available is one with a maximum of 105pF. What do you put in series to reduce the maximum to 56pF?

Did you know the formula above? For two capacitors in series, can become

$$C_{total} = [(C_1 \times C_2)/(C_1 + C_2)]^{**}$$

AND

$$C_{series} = [(C_{avail} \times C_{wanted})/(C_{avail} - C_{wanted})]$$

THEREFORE

$$C_{series} = [(105 \times 56)/(105 - 56)] = 120 pF$$

If you disbelieve me then check it out using ** above as a double check. Nifty eh! Write it in your exercise book. If that flash try this red-hot formula.

Suppose we want an oscillator to tune from 7.0MHz to 7.2MHz. Now if you don't quite know what an oscillator is yet don't worry. We will explore later in the tutorials you will find that your capacitance must vary, for tuning purposes, in the direct ratio of the frequency ratio square or C_{max} to $C_{min} = (F_{high}/F_{low})^2$. Eh what!

In this example our frequency ratio is $F_{high}/F_{low} = 7.2/7.0 = 1.02857$. Our ratio must be squared so then it becomes $(1.02857 \times 1.02857) = 1.05796$. This must be our capacitance ratio i.e. C_{max} to $C_{min} = 1.05796$.

Using the 105pF air variable capacitor is no help because it usually has a minimum of 10.5pF and that's a ratio of 10:1. Assuming the 105pF, for the moment***, is acceptable then apply this formula (which is quite difficult to do in HTML language the way I wanted to do it—so bear with me):

$$(C_{pad} + 105)/(C_{pad} + 10.5) = 1.05796/1$$

What needs to be done is re-write that formula on a piece of paper using two lines and writing the division sign in the usual way, i.e. as an underlining of both $C_{pad} + 105$ and 1.05796. I hope everybody knows what I mean. Next we perform a simple algebraic function, we cross multiply the formula so you should end up with

$$C_{pad} + 105 = (1.05796 \times C_{pad}) + (1.05796 \times 10.5)$$

$$C_{pad} + 105 = (1.05796 \times C_{pad}) + (11.10858)$$

Then subtracting 1 times C_{pad} from each side as well as subtracting 11.10858 from each side (in algebra what is done to one side of the equal sign must be done to the other side—I hope that's clear also) we get

$$93.89142 = 0.05796\ C_{pad}$$ and dividing both sides by 0.05796

$$C_{pad} = 1619.93 pF$$

Now firstly I accept the fact that the formula may be obscure for many people and if that is the case then I will re-write it as a graphic but including too many graphics can cause other problems. It is most important (to me anyway) that I make this as easy as possible for you.

Secondly the figure of 1 619.93pF above is most likely too high to use in the real world. Let's check our sums. We have 105pF max and 10.5pF min in parallel with 1 619.93pF which is the same as (105+1 620)/(10.5+1 620). This calculates out to 1.057 977∶1 which when applied to our tuning range would tune from 7.0MHz to 7.199 99MHz or what we set out to do. Remember our strays earlier. They would play havoc with this, so the 1 620pF would become in the real world 1 000pF + 560pF + 100pF trimmer (i.e. 10~100pF).

I said earlier the 105pF was possibly acceptable ***.

Let's go back and reduce it to say 33pF and redo our sums to get another C_{pad} value. Let me know your answer.

Some home brewers (hi-tech code word for building-it-yourself) advocate using a starting goal of 1pF per wavelength of frequency. e.g. 7.0MHz = 40m (approx.) = 40pF.

I must say I am not especially in love with that particular logic for many reasons which will become apparent as we become more deeply involved with filter and oscillator theory.

New Words and Phrases

capacitor	[kəˈpæsitə]	n.	电容器(= capacitator)
capacitance	[kəˈpæsitəns]	n.	容量,电容
be formed by			由……组成
dielectrics	[ˌdaiiˈlektrik]	n.	电介质,电介体
ceramic	[siˈræmik]	n.	陶瓷,制陶
trimmer	[ˈtrimə]	n.	调整片,微调电容器
inductor	[inˈdʌktə]	n.	电感器,感应器
inherent	[inˈhiərənt]	adj.	固有的,内在的,与生俱来的
stray	[strei]	n.	杂散电容(偶然出现的间层)
resonate	[ˈrezəneit]	n.	共振,共鸣
		vt.	谐振,共鸣,回响,调谐
take…into account			考虑
gotcha	[ˈgɔtʃə]		<口>=(I have) got you
oscillator	[ˈɔsileitə]	n.	振荡器
oscillation	[ˌɔsiˈleiʃən]	n.	振荡,振动
by-pass		n.	支路,旁通
emitter	[iˈmitə]	n.	发射器
rectifier	[ˈrektifaiə]	n.	整流器,矫正器
amplifier	[ˈæmplifaiə]	n.	【电】放大器,扩音器
tutorial	[tjuːˈtɔːriəl]	n.	指南
algebraic	[ˌældʒiˈbreik]	adj.	代数的,关于代数学的
havoc	[ˈhævək]	n.	严重破坏
		vt.	损害
wavelength	[ˈweivleŋθ]	n.	波长

Exercises

【Ex. 6】 根据文章所提供的信息判断正误。

（1）The letter "C" stands for capacitance.

（2）Typical dielectrics you will encounter are air, mica, ceramic or metal.

（3）An air variable capacitor of the type used either as a trimmer or even the tuning element in a radio.

（4）A capacitor will pass DC or RF but not AC. That is often the function of coupling capacitors in circuits.

（5）A reservoir by-pass capacitor used in the output of a DC rectifier to smooth out the power line AC pulses and as a reservoir between the charging pulses.

（6）When three capacitors connected in series, $C_{\text{total}} = C_1 + C_2 + C_3$.

（7）When three capacitors connected in parallel, $C_{\text{total}} = \{1/[(1/C_1) + (1/C_2) + (1/C_n)]\}$.

（8）Stray capacitance in sloppy layouts account for unexpected oscillations, no oscillations, different circuit responses.

（9）Under DC conditions emitter by-pass capacitor sets up how we want the transistor to operate.

（10）Parallel plate capacitors have a formula for calculating capacitance
$$C = 0.22248 \times K \times (A/t)$$

科技英语翻译知识　翻译的标准

翻译标准(Translation norm, Translation criteria, Translation standard principle)是翻译实践的准绳和衡量译文好坏的尺度，也是翻译工作者要努力达到的目标。中外翻译理论家历来认为这是翻译理论的核心问题，因此他们提出了种种论述。

在我国，早在三国时期，支谦就提出了"循本旨，不加文饰"的译经原则。唐代的玄奘提出了"既须求真，又须喻俗"。不过在翻译界影响最大、最强烈的还是清末著名翻译家严复提出的"信、达、雅"三字原则或三字标准。当代的翻译家又有多种说法，有鲁迅的"信、和、顺"标准、朱生豪的"神韵说"、傅雷的"神似说"、钱钟书的"化境说"和许渊冲的"三美说"等。

国外也有多种论述，18世纪的英国学者，爱丁堡大学教授泰勒(A. F. Tytler)提出了翻译三原则：要完整再现原文；保留原文的风格和手法；译文应和原文一样通顺。

前苏联翻译家费道罗夫提出了"等值论"。当代美国翻译家尤金·A·奈达(Eugene. A. Nida)提出了"等效论"，又称动态对等说(Dynamic equivalence)或功能对等论(Functional equivalence)。

国内外的专家对翻译标准有这么多的论述，令初学者感到眼花缭乱，正因为如此，有的专家提出简单明了、普遍能为人们接受的翻译标准，即"忠实通顺"。

科技英语虽有其自身的特点,但是在翻译标准上并不能例外。"忠实通顺"也应当作为科技英语的翻译标准。

1. 忠实

忠实就是指忠于原作的内容,要完整而准确地将原作表达出来,不仅要忠于原作的思想、观点、立场和所流露的感情,还要忠于原作的风格——即原作的民族风格、时代风格和作者个人的语言风格等。总之,原作怎样,译文也应该怎样,尽可能还其本来面目。正如鲁迅所说的,"保存原作的风姿"。

下面通过例句来说明对忠实的理解:

The engine did not stop because the fuel was finished.

译文:发动机没有停止,因为燃料用完了。

这样的译文虽然"完全"忠于原文,但只是忠于原作的形式,并没有忠于原作的内容。译文显然不符合逻辑——这是因为译者缺乏对原文的理解:这是英语中的否定转移,not 并不是修饰谓语动词 stop,而是修饰句子的后面部分。

正确的译文应该是"发动机不是因为燃料用完而停止的"。

又如:

All substances will permit the passage of some electric current, provided the potential difference is high enough.

句中 all substances 相对于 permit 来说是事,但是对于后面的 passage 来说却是地点。因为人们可以将 some electric current, passage 和 all substances 理解为 some electric current pass (through) all substances。这是因为在科技英语中的复合名词词组中,其深层结构中的语义关系比较复杂。如果忽略了这一点,只按照表层结构理解,译文就是:"只要有足够的电位差,所有的物体都容许一些电流通过"。如果从语义关系来考虑,译文应是:"只要有足够的电位差,电流便可以通过任何物体"。

可见忠实并不仅仅是忠于文字的表面,而是忠于语言的内容。

2. 通顺

通顺就是指译文必须通顺易懂,符合规范。具体地说,科技英语的翻译要符合科技语言的规范,要用明白晓畅的现代语言,没有逐词死译、硬译的现象,没有语言晦涩、文理不通、结构混乱、逻辑不清的现象。下面通过实例来说明对通顺的理解。

The virtual reality technology is hindered right now by the fact that today's computers are simply not fast enough.

译文:虚拟现实技术被今天计算机不够快速所制约。

译者试图尽量对原文忠实翻译,但是这样的译文读起来非常别扭,原因是没有注意到英汉语言在表达上的差异,所以一点儿也不通顺。

正确的译文应该是:

目前,计算机运行速度缓慢制约着虚拟现实技术的发展。

这个译文对原文的次序做了调整,还增加了原文中形无而实有的"运行速度"一词;把原文的被动语态改为汉语中的主动语态。此译文不仅忠于原文的内容,也符合汉语的规范。

忠实和通顺是相辅相成的。忠实而不通顺,读者看不懂,也就谈不到忠实;通顺而不忠实,也就使译文失去了原意,成为杜撰。

所以,在科技英语的翻译中,以"忠实通顺"作为翻译标准是切实可行的。

Reading Material

阅读下列文章。

Text	Note
Inductor An inductor[1], also called a coil[2] or reactor[3], is a passive two-terminal electrical component that stores electrical energy in a magnetic field when electric current is flowing through it. When the current flowing through an inductor changes, the time-varying magnetic field[4] induces a voltage in the conductor, described by Faraday's law[5] of induction. According to Lenz's law[6], the direction of induced electromotive force (e. m. f.)[7] opposes the change in current that created it. As a result, inductors oppose any changes in current. An inductor typically consists of an electric conductor, such as a wire, that is wound into a coil. An inductor is characterized by its inductance, which is the ratio of the voltage to the rate of change of current. In the International System of Units (SI), the unit of inductance is the henry (H). Inductors have values that typically range from 1 μH (10^{-6}H) to 1 H. Many inductors have a magnetic core[8] made of iron or ferrite inside the coil, which serves to increase the magnetic field and thus the inductance[9]. Along with capacitors and resistors, inductors are one of the three passive linear[10] circuit elements that make up electric circuits. Inductors are widely used in alternating current (AC) electronic equipment, particularly in radio equipment. They are used to block AC while allowing DC to pass; inductors designed for this purpose are called chokes[11]. They are also used in electronic filters[12] to separate signals of different frequencies, and in combination with capacitors to make tuned circuits, used to tune radio and TV receivers. **1. Overview** Inductance (L) results from the magnetic field around a current-carrying conductor; the electric current through the conductor creates a magnetic flux[13]. Mathematically speaking, inductance is determined by how much magnetic flux φ through the circuit is created by a given current i $$L = \varphi/i \quad (1)$$ Inductors that have ferromagnetic cores are nonlinear[14]; the inductance changes with the current, in this more general case inductance is defined as $$L = d\varphi/di$$ Any wire or other conductor will generate a magnetic field when current flows through it, so every conductor has some inductance. The inductance of a circuit depends	[1] *n.* 电感应器 [2] *n.* 线圈 [3] *n.* 反应器,电抗器 [4] magnetic field: 磁场 [5] Faraday's law: 法拉第定律 [6] Lenz's law: 愣次定律 [7] induced electromotive force: 感应电动势 [8] magnetic core: 磁芯 [9] *n.* 感应系数,自感应 [10] *adj.* 线性的 [11] *n.* 电感扼流圈 [12] *n.* 滤波器,过滤器 [13] magnetic flux: 磁通量 [14] *adj.* 非线性的

on the geometry[15] of the current path as well as the magnetic permeability[16] of nearby materials. An inductor is a component consisting of a wire or other conductor shaped to increase the magnetic flux through the circuit, usually in the shape of[17] a coil or helix. Winding the wire into a coil increases the number of times the magnetic flux lines link the circuit, increasing the field and thus the inductance. The more turns, the higher the inductance. The inductance also depends on the shape of the coil, separation of the turns, and many other factors[18]. By adding a "magnetic core" made of a ferromagnetic material like iron inside the coil, the magnetizing field from the coil will induce magnetization in the material, increasing the magnetic flux. The high permeability[19] of a ferromagnetic core can increase the inductance of a coil by a factor of several thousand over what it would be without it.

1.1 Constitutive equation

Any change in the current through an inductor creates a changing flux, inducing a voltage across the inductor. By Faraday's law of induction, the voltage induced by any change in magnetic flux through the circuit is

$$v = d\varphi/dt$$

From (1) above

$$v = L(di/dt) \qquad (2)$$

So inductance is also a measure of the amount of electromotive force[20] (voltage) generated for a given rate of change of current. For example, an inductor with an inductance of 1 henry produces an EMF of 1 volt when the current through the inductor changes at the rate of 1 Ampere[21] per second. This is usually taken to be the constitutive[22] relation (defining equation[23]) of the inductor.

The dual of the inductor is the capacitor, which stores energy in an electric field rather than a magnetic field. Its current-voltage relation is obtained by exchanging current and voltage in the inductor equations and replacing L with the capacitance[24] C.

1.2 Lenz's law

The polarity (direction) of the induced voltage is given by Lenz's law, which states that it will be such as to oppose the change in current. For example, if the current through an inductor is increasing, the induced voltage[25] will be positive at the terminal through which the current enters and negative at the terminal through which it leaves, tending to oppose the additional current. The energy from the external circuit necessary to overcome this potential "hill" is being stored in the magnetic field of the inductor; the inductor is said to be "charging" or "energizing". If the current is decreasing, the induced voltage will be negative at the terminal through which the current enters and positive at the terminal through which it leaves, tending to maintain the current. Energy from the magnetic field is being returned to the circuit; the inductor is said to be "discharging[26]".

1.3 Ideal and real inductors

In circuit theory, inductors are idealized as obeying the mathematical relation (2) above precisely. An "ideal inductor" has inductance, but no resistance or capacitance, and does not dissipate or radiate energy. However, real inductors have

[15] n. 几何形状
[16] magnetic permeability: 导磁率
[17] in the shape of: 以……的形式
[18] n. 因素, 要素, 因数
[19] n. 渗透性
[20] electromotive force: 电动势
[21] n. 安培
[22] adj. 构成的, 制定的
[23] n. 方程式, 等式
[24] n. 电容
[25] vt. 感应电压
[26] adj. 放电的

side effects which cause their behavior to depart from[27] this simple model. They have resistance (due to the resistance of the wire and energy losses in core material), and parasitic capacitance[28] (due to the electric field between the turns of wire which are at slightly different potentials). At high frequencies the capacitance begins to affect the inductor's behavior; at some frequency, real inductors behave as resonant circuits[29], becoming self-resonant[30]. Above the resonant frequency the capacitive reactance becomes the dominant part of the impedance. At higher frequencies, resistive losses in the windings increase due to skin effect[31] and proximity effect[32].

Inductors with ferromagnetic cores have additional energy losses due to hysteresis[33] and eddy currents[34] in the core, which increase with frequency. At high currents, iron core inductors also show gradual departure from ideal behavior due to nonlinearity caused by magnetic saturation[35] of the core. An inductor may radiate electromagnetic energy into surrounding space and circuits, and may absorb electromagnetic emissions from other circuits, causing electromagnetic interference (EMI[36]). Real-world inductor applications may consider these parasitic parameters[37] as important as the inductance.

2. Applications

Inductors are used extensively in analog circuits and signal processing[38]. Applications range from the use of large inductors in power supplies to the small inductance of the ferrite bead or torus installed around a cable to prevent radio frequency interference[39] from being transmitted down the wire. Inductors are used as the energy storage device in many switched-mode power supplies to produce DC current. The inductor supplies energy to the circuit to keep current flowing during the "off" switching periods.

An inductor connected to a capacitor forms a tuned circuit, which acts as a resonator for oscillating current[40]. Tuned circuits are widely used in radio frequency equipment such as radio transmitters and receivers, as narrow bandpass filters[41] to select a single frequency from a composite signal, and in electronic oscillators[42] to generate sinusoidal[43] signals.

Two (or more) inductors in proximity that have coupled magnetic flux (mutual inductance[44]) form a transformer[45], which is a fundamental component of every electric utility power grid. The efficiency of a transformer may decrease as the frequency increases due to eddy currents in the core material and skin effect on the windings. The size of the core can be decreased at higher frequencies. For this reason, aircraft use 400 hertz alternating current rather than the usual 50 or 60 hertz, allowing a great saving in weight from the use of smaller transformers.

Inductors are also employed in electrical transmission systems, where they are used to limit switching currents and fault currents. In this field, they are more commonly referred to as reactors[46].

Because inductors have complicated side effects which cause them to depart from ideal behavior, because they can radiate electromagnetic interference (EMI), and most of all, because of their bulk which prevents them from being integrated on semiconductor chips[47], the use of inductors is declining in modern electronic devices,

[27] *n.* depart from: 离开
[28] parasitic capacitance: 寄生电容
[29] resonant circuits: 谐振电路
[30] self-resonant: 自谐振
[31] skin effect: 趋肤效应
[32] proximity effect: 邻近效应
[33] *n.* 磁滞现象
[34] eddy currents: 涡电流
[35] magnetic saturation: 磁饱和
[36] EMI: 电磁干扰
[37] parasitic parameters: 寄生参数
[38] signal processing: 信号处理
[39] *n.* 冲突, 干涉
[40] oscillating current: 振荡电流
[41] narrow bandpass filters: 窄带滤波器
[42] electronic oscillators: 电子振荡器
[43] *n.* 正弦曲线
[44] mutual inductance: 互感(系数)
[45] *n.* 变压器
[46] *n.* 电抗器
[47] semiconductor chip: 半导体芯片

particularly compact portable devices. Real inductors are increasingly being replaced by active circuits such as the gyrator[48] which can synthesize inductance using capacitors.

3. Inductor construction

An inductor usually consists of a coil of conducting material, typically insulated[49] copper wire, wrapped around a core either of plastic[50] or of a ferromagnetic[51] (or ferrimagnetic[52]) material; the latter is called an "iron core" inductor. The high permeability of the ferromagnetic core increases the magnetic field and confines it closely to the inductor, thereby increasing the inductance. Low frequency inductors are constructed like transformers, with cores of electrical steellaminated to prevent eddy currents. "Soft" ferrites are widely used for cores above audio frequencies, since they do not cause the large energy losses at high frequencies that ordinary iron alloys do. Inductors come in many shapes. Most are constructed as enamel coated wire (magnet[53] wire) wrapped around a ferrite bobbin[54] with wire exposed on the outside, while some enclose[55] the wire completely in ferrite and are referred to as "shielded[56]". Some inductors have an adjustable core, which enables changing of the inductance. Inductors used to block very high frequencies are sometimes made by stringing a ferrite bead on a wire.

Small inductors can be etched[57] directly onto a printed circuit board[58] by laying out the trace in a spiral pattern. Some such planar inductors use a planar core.

Small value inductors can also be built on integrated circuits using the same processes that are used to make transistors[59]. Aluminium[60] interconnect is typically used, laid out in a spiral coil pattern. However, the small dimensions limit the inductance, and it is far more common to use a circuit called a "gyrator" that uses a capacitor and active components to behave similarly to an inductor.

[48] *n.* 回转器
[49] *n.* 绝缘的
[50] *n.* 塑胶,塑料
[51] *adj.* 铁磁的
[52] *adj.* 亚铁磁的
[53] *n.* 磁体,磁铁
[54] *n.* 线轴,绕线筒
[55] *vt.* 封装
[56] *adj.* 防护的,铠装的,屏蔽了的,隔离的
[57] *v.* 蚀刻
[58] printed circuit board: 印制电路板
[59] *n.* 晶体管
[60] *n.* 铝 *adj.* 铝的

参考译文 电阻和电导

电导体的电阻是电流通过该导体的难度的度量。其反向量是电导,表示电流通过的容易性。在概念上,电阻与机械摩擦有些相似。电阻的 SI 单位是欧姆(Ω),而电导以西门子(S)来度量。

具有均匀横截面的物体的电阻与其电阻率和长度成正比,且与其横截面积成反比。除了具有零电阻的超导体以外,所有材料都有一定量的电阻。

物体的电阻(R)被定义为其上的电压(V)与通过其的电流(I)的比率,而电导(G)是相反的:

$$R = V/I \quad G = I/V = 1/R$$

对于多种多样的材料和条件,V 和 I 彼此成正比,因此 R 和 G 是恒定的(尽管它们可以随如温度这样的其他因素而变化)。该比例被称为欧姆定律,并且满足该定律的材料被称为欧姆材料。

在其他情况下,例如二极管或电池,V 和 I 不成正比。比率 V/I 有时仍然有用,并且被称为"静态电阻",因为它对应于原点和 $I-V$ 曲线之间的弦的反斜率。在其他情况下,导数

dV/dI 可能非常有用,这被称为"差分电阻"。

1. 说明

(图略)

将流经电路的电流与流过管道的水进行比较。当管(左)中充满头发(右)时,需要更大的压力来实现相同的水流。电流要流过大电阻就像水流过被毛发堵塞的管道:需要更大的推力(电动势)来驱动相同的流动(电流)。

在液压模拟中,流过导线(或电阻器)的电流类似于流经管道的水,穿过导线的电压降类似于推动水通过管道的压力降。电导与给定压力下的流量成正比,电阻与实现给定流量所需的压力成正比。(电导和电阻互为倒数。)

电压降(例如,电阻器两侧电压之间的差)——而不是电压本身——提供推动电流通过电阻器的驱动力。这与液压类似:管道两侧之间的压力差——而不是压力本身——决定了通过它的流量。例如,在管道上方可能存在大的水压,试图将水向下推过管道。但是在管道下方可能存在相同大的水压,试图将水通过管道推回。如果这两个压力相等,水就不会流动。

电线、电阻器或其他元件的电阻和电导主要由两个属性决定:

- 几何(形状)
- 材料

几何形状很重要,因为要让水通过长而窄的管道比通过宽而短的管道更困难。同样,长而细的铜线比短而粗的铜线有更高的电阻(更低的电导)。

材料也很重要。填充头发的管子比具有相同形状和尺寸的清洁管子通过水流更少。类似地,电子可以自由且容易地流过铜线,但是不易流过具有相同形状和尺寸的钢丝,并且它们基本上不能流过绝缘体(如橡胶),无论这些绝缘体形状如何。铜、钢和橡胶之间的差异与它们的微观结构和电子构型有关,并且通过被称为电阻率的特性来量化。

2. 导体和电阻器

一个 6.5MΩ 电阻,由其电子颜色代码(蓝—绿—黑—黄)表示。可以使用欧姆表来检验该值。

(图略)

电可以流过的物体被称为导体。电路中使用的有特定阻力的导电材料称为电阻器。导体由高导电性材料(例如金属,特别是铜和铝)制成。另一方面,电阻器由多种材料制成,这取决于期望的电阻、需要消耗的能量、精度和成本等因素。

3. 电阻率和电导率的关系

给定物体的阻力主要取决于两个因素:其材料以及形状。对于给定的材料,电阻与横截面积成反比;例如,在其他方面相同的条件下,粗铜线的电阻比细铜线的电阻低。此外,对于给定材料,电阻与长度成正比;例如,在其他方面相同的条件下,长铜线比短铜线的电阻高。因此,均匀横截面导体的电阻 R 和电导 G 可以计算为:

$$R = \rho(L/A)$$
$$G = \sigma(A/L)$$

其中，L 是导体长度，以米（m）度量；A 是导体的横截面积，以平方米（m^2）度量；σ（sigma）是电导率，以西门子/米（$S \cdot m^{-1}$）度量，ρ 是材料的电阻率（也称为比电阻），以欧姆·米（$\Omega \cdot m$）度量。电阻率和电导率是比例常数，因此仅取决于线材的材料，而不是线材的几何形状。电阻率和电导率互为倒数：$\rho = 1/\sigma$。电阻率量度材料抵抗电流的能力。

该公式不精确，因为它假定导体中的电流密度完全均匀，在实际情况中这并不总是真实的。然而，对于像导线这样的长细导体，该公式仍然最符合实际。

该公式不精确的另一种情况是使用于交流电（AC）时，因为趋肤效应抑制了导体中心附近的电流流动。因此，几何横截面不同于电流实际流过的有效横截面，所以电阻比预期的要高。类似地，如果承载 AC 电流的两个导体彼此相邻，则由于邻近效应，它们的电阻会增加。在商业电源频率下，这些效应对于承载大电流的大导体（例如，变电站中的母线或承载大于几百安培的大功率电缆）是显著的。

4. 测量电阻

用于测量电阻的仪器称为欧姆表。简单的欧姆表不能准确地测量低电阻，因为它们的测量引线的电阻导致干扰测量的电压降，因此更精确的器件使用四端子检测技术。

Unit 3

Text A Simple Electrical Circuit

1. An electric circuit

A fundamental relationship exists between current, voltage, and resistance. A simple electrical circuit consists of a voltage source, some type of load, and a conductor to allow electrons to flow between the voltage source and the load. [1] In the following circuit a battery provides the voltage source, electrical wire is used for the conductor, and a light provides the resistance. An additional component has been added to this circuit, a switch. There must be a complete path for current to flow. If the switch is open, the path is incomplete and the light will not illuminate. Closing the switch completes the path, allowing electrons to leave the negative terminal and flow through the light to the positive terminal.

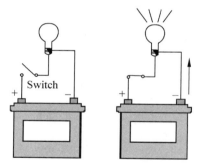

2. An electrical circuit schematic

The following schematic is a representation of an electrical circuit, consisting of a battery, a resistor, a voltmeter and an ammeter. The ammeter, connected in series with the circuit, will show how much current flows in the circuit. The voltmeter, connected across the voltage source, will show the value of voltage supplied from the battery. Before an analysis can be made of a circuit, we need to understand Ohm's Law.

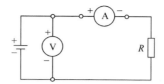

3. Ohm's Law

The relationship between current, voltage and resistance was studied by the 19th century German mathematician, George Simon Ohm. Ohm formulated a law which states that current varies proportionally with voltage and inversely with resistance. From this law the following formula is derived:

$$I = \frac{U}{R} \quad \text{or} \quad \text{Current} = \frac{\text{Voltage}}{\text{Resistance}}$$

Ohm's Law is the basic formula used in all electrical circuits. Electrical designers must decide how much voltage is needed for a given load, such as computers, clocks, lamps and motors. Decisions must be made concerning the relationship of current, voltage and resistance. All electrical design and analysis begins with Ohm's Law. There are three mathematical ways to express Ohm's Law. Which of the formulas is used depends on what facts are known before starting and what facts need to be known.

$$I = \frac{U}{R} \quad U = I \times R \quad R = \frac{U}{I}$$

4. Ohm's Law triangle

There is an easy way to remember which formula to use. By arranging current, voltage and resistance in a triangle, one can quickly determine the correct formula.

5. Using the triangle

To use the triangle, cover the value you want to calculate. The remaining letters make up the formula. [2]

Ohm's Law can only give the correct answer when the correct values are used. Remember the following three rules:
- Current is always expressed in amperes or amp.
- Voltage is always expressed in volt.
- Resistance is always expressed in ohm.

DC Series Circuit

1. Resistance in a series circuit

A series circuit is formed when a number of resistors are connected end-to-end so that there is only one path for current to flow. [3] The resistors can be actual resistors or other devices that have resistance. The illustration shows four resistors connected end-to-end. There is one path of electron flow from the negative terminal of the battery through R_4, R_3, R_2, R_1 returning to the positive terminal.

2. Formula for series resistance

The values of resistance add in a series circuit. If a 4Ω resistor is placed in series with a 6Ω resistor, the total value will be 10Ω. This is true when other types of resistive devices are placed in series. The mathematical formula for resistance in series is

$$R_t = R_1 + R_2 + R_3 + R_4$$

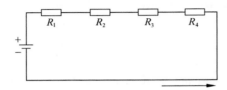

Given a series circuit where R_1 is $11k\Omega$, R_2 is $2k\Omega$, R_3 is $2k\Omega$, R_4 is 100Ω, and R_5 is $1k\Omega$, what is the total resistance?

$$\begin{aligned} R_t &= R_1 + R_2 + R_3 + R_4 + R_5 \\ &= 11\,000 + 2\,000 + 2\,000 + 100 + 1\,000 \\ &= 16\,100\Omega \end{aligned}$$

3. Current in a series circuit

The equation for total resistance in a series circuit allows us to simplify a circuit. Using Ohm's Law, the value of current can be calculated. Current is the same anywhere when it is measured in a series circuit.

$$I = \frac{U}{R} = \frac{12}{10} = 1.2A$$

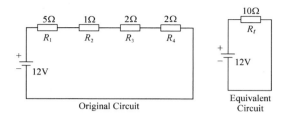

Original Circuit   Equivalent Circuit

4. Voltage in a series circuit

Voltage can be measured across each of the resistors in a circuit. The voltage across a resistor is referred to as a voltage drop. A German physicist, Kirchhoff, formulated a law which states the sum of the voltage drops across the resistances of a closed circuit equals the total voltage applied to the circuit.[4] In the following illustration, four equal value resistors of 1.5Ω each have been placed in series with a 12V battery. Ohm's Law can be applied to show that each resistor will "drop" an equal amount of voltage.

First, solve for total resistance:

$$R_t = R_1 + R_2 + R_3 + R_4 = 1.5 + 1.5 + 1.5 + 1.5 = 6\Omega$$

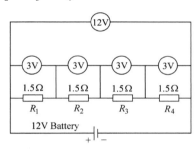

Second, solve for current:

$$I = \frac{U}{R} = \frac{12}{6} = 2\text{A}$$

Third, solve for voltage across any resistor:

$$U = I \times R = 2 \times 1.5 = 3\text{V}$$

If voltages were measured across any single resistor, the voltmeter would read 3V.[5] If voltage were measured across a combination of R_3 and R_4 the voltmeter would read 6V. If voltage were measured across a combination of R_2, R_3, and R_4 the voltmeter would read 9V. If the voltage drops of all four resistors were added together the sum would be 12V, the original supply voltage of the battery.

5. Voltage division in a series circuit

It is often desirable to use a voltage potential that is lower than the supply voltage. To do this, a voltage divider can be used. The battery represents U_1 which in this case is 50V. The desired voltage is represented by U_0 which mathematically works out to be 40V. To calculate this voltage, first solve for total resistance:

$$R_t = R_1 + R_2 = 5 + 20 = 25\Omega$$

Second, solve for current:

$$I = \frac{U_1}{R_t} = \frac{50}{25} = 2\text{A}$$

Finally, solve for voltage:

$$U_O = I \times R_2 = 2 \times 20 = 40\text{V}$$

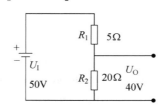

New Words and Phrases

circuit	[ˈsə:kit]	n. 电路,一圈,周游,巡回
fundamental	[ˌfʌndəˈmentl]	adj. 基础的,基本的
consist of		由……组成
load	[ləud]	n. 负荷,负载,加载
battery	[ˈbætəri]	n. 电池
component	[kəmˈpəunənt]	n. 成分
switch	[switʃ]	n. 开关,电闸,转换
illuminate	[iˈlju:mineit]	vt. 阐明,说明(问题等),照明,照亮
negative	[ˈnegətiv]	n. 否定,负数
		adj. 否定的,消极的,负的,阴性的
		vt. 否定,拒绝(接受)
positive	[ˈpɔzətiv]	adj. 阳的
		adj. 肯定的,积极的,绝对的,确实的
		adj. [数]正的
schematic	[skiˈmætik]	adj. 示意性的
		n. 电路原理图
voltmeter	[ˈvəultˌmi:tə(r)]	n. 电压表
ammeter	[ˈæmitə]	n. 电流表
Ohm's Law		欧姆定律
mathematician	[ˌmæθiməˈtiʃən]	n. 数学家
vary	[ˈvɛəri]	vt. 改变,变更,使多样化
		vi. 变化,不同,违反
inverse	[ˈinˈvə:s]	adv. 相反地,倒转地
derive	[diˈraiv]	vt. 得自
		vi. 起源

designer	[dɪˈzaɪnə(r)]	n. 设计者
lamp	[læmp]	n. 灯
relationship	[rɪˈleɪʃənʃɪp]	n. 关系,关联
analysis	[əˈnælɪsɪs]	n. 分析,分解
triangle	[ˈtraɪæŋgl]	n. 三角形
calculate	[ˈkælkjuleɪt]	vt. & vi. 计算,考虑,计划,打算
		vt. & vi. (美) 以为,认为
series circuit		串联电路
equation	[ɪˈkweɪʃən]	n. 相等,平衡,综合体,因素,方程式,等式
series resistance		串联电阻
be referable to		可归因于,与……有关
voltage drop		电压降
meter	[ˈmiːtə]	n. 仪表,米,计,表
divider	[dɪˈvaɪdə]	n. 分割者,间隔物,分配器

Notes

[1] A simple electric circuit consists of a voltage source, some type of load, and a conductor to allow electrons to flow between the voltage source and the load.

本句中的谓语动词是 consist of,意为"由……组成"。to allow electrons to flow between the voltage source and the load 修饰的是 conductor,表明是什么样的导线,而不是整个句子。

[2] To use the triangle, cover the value you want to calculate. The remaining letters make up the formula.

这两个句子关系紧密,要联系起来理解。后一个句子表明的是使用三角形,盖住要计算的值的结果,剩下的字母组成公式。

[3] A series circuit is formed when a number of resistors are connected end-to-end so that there is only one path for current to flow.

本句中的 end-to-end 不能凭字面理解为尾对尾,而是首尾相连的意思。so that 引导了一个结果状语从句。when 引导的从句作状语,指明串联电路形成的条件。

[4] A German physicist, Kirchhoff, formulated a law which states the sum of the voltage drops across the resistances of a closed circuit equals the total voltage applied to the circuit.

看懂这个句子的关键是分析它的句子结构。这是一个多层从句的句子。全句的主语是 A German physicist,谓语是 formulated,宾语是 a law,Kirchhoff 是同位语。which 引导的定语从句修饰 a law。在该定语从句中,which 作主语,states 是谓语动词,states 后又是一个宾语从句,省略了引导词 that。在这个宾语从句中,主语为 the sum of the voltage drops,谓语为 equals,宾语为 the total voltage。结构清楚后,整个句子的意思就一目了然了:德国物理学家陈述了一条定律,定律的内容是,穿过闭路电阻的电压降的总和等于这个回路上的电压。

[5] If voltage were measured across any single resistor, the voltmeter would read 3V.

注意,"表的读数为……"的表达是本句中的 volt meter would read,而不是 voltmeter would be read。read 应理解"显示,指示"。例如,The dial reads 32. 刻度显示出 32。

Exercises

【Ex. 1】 根据课文内容,回答以下问题。

(1) What does Ohm's Law state?

(2) According to the passage, how to use the triangle?

(3) What is a series circuit?

(4) How do we measure the voltage drop of each of the resistors in a circuit?

(5) If three resistors of 10Ω, 20Ω and 30Ω respectively have been placed in series with a 12V battery, what is the voltage drop of each of the resistors in a circuit?

【Ex. 2】 根据下面的英文解释,写出相应的英文词汇。

英 文 解 释	词 汇
a closed path followed or capable of being followed by an electric current	
a device used to break or open an electric circuit or to divert current from one conductor to another	
a position in a circuit or device at which a connection is normally established or broken	
an instrument, such as galvanometer, for measuring potential differences in volts	
an instrument that measures electric current	
a device that generates light, heat, or therapeutic radiation	
a device that converts any form of energy into mechanical energy, especially an internal-combustion engine or an arrangement of coils and magnets that converts electric current into mechanical power	
scientist who specializes in physics	
the work required to bring a unit electric charge, magnetic pole, or mass from an infinitely distant position to a designated point in a static electric, magnetic, or gravitational field, respectively	
the international standard unit of length, approximately equivalent to 39.37 inches. It was redefined in 1983 as the distance traveled by light in a vacuum in 1/299 792 458 second	

【Ex.3】 把下列句子翻译为中文。

(1) A power supply could be something as simple as a 9V battery or it could be as complex as a precision laboratory power supply.

(2) Variable resistors are common components. They have a dial or a knob that allows you to change the resistance. This is very useful for many situations.

(3) Diodes are components that allow current to flow in only one direction. They have a positive side and a negative side.

(4) LEDs use a special material which emits light when current flows through it. Unlike light bulbs, LEDs never burn out unless their current limit is reached.

(5) Well the letter L stands for inductance. The simplest inductor is consists of a piece of wire.

(6) Two metallic plates separated by a non-conducting material between them make a simple capacitor.

(7) The time required for a capacitor to reach its maximum charge is proportional to the capacitance value and the resistance value.

(8) When AC current flows through an inductance a opposite emf or voltage develops opposing any change in the initial current.

(9) Reactance is the property of resisting or impeding the flow of AC current or AC voltage in inductors and capacitors.

(10) To produce a drift of electrons, or electric current, along a wire it is necessary that there be a difference in "pressure" or potential between the two ends of the wire. This potential

difference can be produced by connecting a source of electrical potential to the ends of the wire.

【Ex. 4】 把下列短文翻译成中文。

Switches are devices that create a short circuit or an open circuit depending on the state of the switch. For a light switch, ON means short circuit (current flows through the switch, lights light up). When the switch is OFF, that means there is an open circuit (no current flows, lights go out). When the switch is ON it looks and acts like a wire. When the switch is OFF there is no connection.

【Ex. 5】 通过 Internet 查找资料，借助如"金山词霸"等电子词典和辅助翻译软件，完成以下技术报告。通过 E-mail 发送给老师，并附上你收集资料的网址。

(1) 一个电路包括哪些主要元件，各种元件由哪些公司生产（附上各种最新产品的图片）。

(2) 叙述德国物理学家基尔霍夫的生平简历及其重大贡献。

Text B DC Parallel Circuit

1. Resistance in a parallel circuit

A parallel circuit is formed when two or more resistances are placed in a circuit side-by-side so that current can flow through more than one path. The illustration shows two resistors placed side-by-side. There are two paths of current flow. One path is from the negative terminal of the battery through R_1 returning to the positive terminal. The second path is from the negative terminal of the battery through R_2 returning to the positive terminal of the battery.

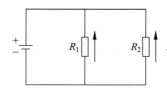

2. Formula for equal value resistors in a parallel circuit

To determine the total resistance when resistors are of equal value in a parallel circuit, use the following formula:

$$R_t = \frac{\text{Value of one resistor}}{\text{Number of resistors}}$$

In the following illustration there are three 15Ω resistors. The total resistance is

$$R_t = \frac{\text{Value of one resistor}}{\text{Number of resistors}} = \frac{15}{3} = 5\Omega$$

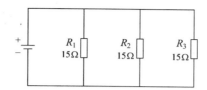

3. Formula for unequal resistors in a parallel circuit

There are two formulas to determine total resistance for unequal value resistors in a parallel circuit. The first formula is used when there are three or more resistors. The formula can be extended for any number of resistors.

$$\frac{1}{R_t} = \frac{1}{R_1} + \frac{1}{R_2} + \frac{1}{R_3}$$

In the following illustration there are three resistors, each has different value. The total resistance is

$$\frac{1}{R_t} = \frac{1}{R_1} + \frac{1}{R_2} + \frac{1}{R_3}$$

$\frac{1}{R_t} = \frac{1}{5} + \frac{1}{10} + \frac{1}{20}$ Insert value of the resistors

$= \frac{4}{20} + \frac{2}{20} + \frac{1}{20}$ Find lowest common multiple

$= \frac{7}{20}$ Add the numerators

$\frac{R_t}{1} = \frac{20}{7}$ Invert both sides of the equation

$R_t = 2.86\Omega$ Divide

The second formula is used when there are only two resistors.

$$R_t = \frac{R_1 \times R_2}{R_1 + R_2}$$

In the following illustration there are two resistors, each has different value. The total resistance is

$$R_t = \frac{R_1 \times R_2}{R_1 + R_2} = \frac{5 \times 10}{5 + 10}$$

$$= \frac{50}{15} \approx 3.33\Omega$$

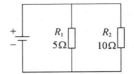

4. Voltage in a parallel circuit

When resistors are placed in parallel across a voltage source, the voltage is the same across each resistor. In the following illustration three resistors are placed in parallel across a 12V battery. Each resistor has 12V available to it.

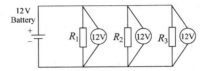

5. Current in a parallel circuit

Current flowing through a parallel circuit divides and flows through each branch of the circuit.

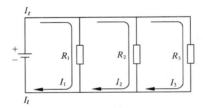

Total current in a parallel circuit is equal to the sum of the current in each branch. The following formula applies to current in a parallel circuit

$$I_t = I_1 + I_2 + I_3$$

6. Current flow with equal value resistors in a parallel circuit

When equal resistances are placed in a parallel circuit, current flow is the same in each branch. In the following circuit R_1 and R_2 are of equal value. If total current (I_t) is 10A, then 5A would flow through R_1 and 5A would flow through R_2.

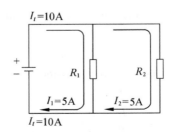

$$I_t = I_1 + I_2 = 5A + 5A = 10A$$

7. Current flow with unequal value resistors in a parallel circuit

When unequal value resistors are placed in a parallel circuit, current flow is not the same in every circuit branch. Current is greater through the path of least resistance. In the following circuit R_1 is 40Ω and R_2 is 20Ω. Small value of resistance means less current flow. More current will flow through R_2 than R_1.

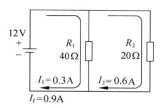

Using Ohm's Law, the total current for the circuit can be calculated.

$$I_1 = \frac{U}{R_1} = \frac{12\text{V}}{40\Omega} = 0.3\text{A}$$

$$I_2 = \frac{U}{R_2} = \frac{12\text{V}}{20\Omega} = 0.6\text{A}$$

$$I_t = I_1 + I_2 = 0.3\text{A} + 0.6\text{A} = 0.9\text{A}$$

Total current can also be calculated by the first calculating total resistance, then applying the formula for Ohm's Law.

$$R_t = \frac{R_1 \times R_2}{R_1 + R_2} = \frac{40\Omega \times 20\Omega}{40\Omega + 20\Omega} = \frac{800\Omega^2}{60\Omega} \approx 13.333\Omega$$

$$I_t = \frac{U}{R_t} = \frac{12\text{V}}{13.333\Omega} \approx 0.9\text{A}$$

8. Series-parallel circuit

Series-parallel circuit is also known as compound circuit. At least three resistors are required to form a series-parallel circuit. The following illustrations show two ways a series-parallel circuit could be formed.

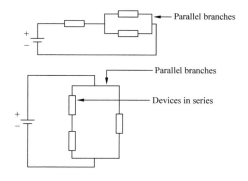

9. Simplifying a series-parallel circuit to a series circuit

The formulas required for solving current, voltage and resistance problems have already been defined. To solve series-parallel circuit, simplify the compound circuits to equivalent simple circuits. In the following illustration R_1 and R_2 are parallel with each other. R_3 is in series with the parallel circuit of R_1 and R_2.

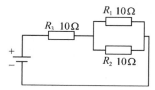

First, use the formula to determine total resistance of a parallel circuit to find the equivalent resistance of R_1 and R_2. When the resistors in a parallel circuit are equal, the following formula is used:

$$R = \frac{\text{Value of any one resistor}}{\text{Number of resistors}} = \frac{10\Omega}{2} = 5\Omega$$

Second, redraw the circuit showing the equivalent values. The result is a simple series circuit which uses already learned equations and methods of problem solving.

10. Simplifying a series-parallel circuit to a parallel circuit

In the following illustration R_1 and R_2 are in series with each other. R_3 is in parallel with the series circuit of R_1 and R_2.

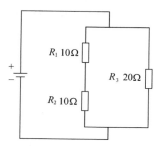

First, use the formula to determine total resistance of the series circuit to find the total resistance of R_1 and R_2. The following formula is used:

$$R = R_1 + R_2 = 10\Omega + 10\Omega = 20\Omega$$

Second, redraw the circuit showing the equivalent values. The result is a simple parallel circuit which uses already learned equations and methods of problem solving.

New Words and Phrases

parallel circuit		并联电路
side-by-side		*adj.* 并肩的,并行的
equal value resistors		等值电阻
multiple	['mʌltipl]	*n.* 倍数,若干
lowest common multiple		最小公倍数
numerator	['njuːməreitə]	*n.* 分子
invert	[in'vəːt]	*adj.* 转化的
		vt. 使颠倒,使转化
		n. 颠倒的事物
amp	[æmp]	*n.* 安培
branch	[brɑːntʃ]	*n.* 枝,分支,支流,支脉
apply to		将……应用于
flow through		流过
series-parallel		串-并联
compound	['kɔmpaund]	*n.* 混合物,[化]化合物
		adj. 复合的
		vt. & vi. 混合,配合
compound circuits		复合电路
parallel branch		并联分支
simplify	['simplifai]	*vt.* 单一化,简单化
reduce	[ri'djuːs]	*vt.* 减少,缩小,简化,还原
equivalent	[i'kwivələnt]	*adj.* 相等的,相当的,同意义的
		n. 等价物,相等物
method	['meθəd]	*n.* 方法
redraw	[riː'drɔː]	*vt.* 重画
		vi. 刷新(屏幕)

Exercises

【Ex. 6】 根据文章所提供的信息判断正误。

(1) A parallel circuit is formed when two or more resistances are placed in a circuit side by-side so that current can flow through only one path.

(2) To determine the equivalent resistance when resistors are of equal value in a parallel circuit, use the following formula

$$R_t = \frac{\text{Value of one resistor}}{\text{Number of resistors}}$$

(3) In the following illustration there are three 15Ω resistors. The equivalent resistance is 45Ω.

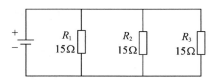

(4) In the following illustration there are three resistors, each has different value. The equivalent resistance is

$$\frac{1}{R_t} = \frac{1}{R_1} + \frac{1}{R_2} + \frac{1}{R_3}$$

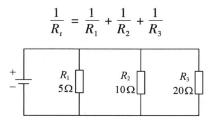

(5) When resistors are placed in series across a voltage source, the voltage is the same across each resistor.

(6) Current flowing through a parallel circuit divides and flows through each branch of the circuit. Total current in a parallel circuit is equal to the sum of the current in each branch.

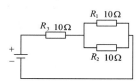

(7) When different resistances are placed in a parallel circuit, current flow is the same in each branch.

(8) Series-parallel circuit is also known as compound circuits. At least more than two resistors are required to form a series-parallel circuit.

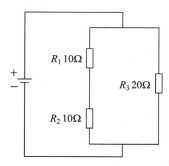

(9) In the following illustration R_1 and R_2 are series with each other. R_3 is in parallel with the series circuit of R_1 and R_2.

(10) In the following illustration, the total resistance is 105Ω.

科技英语翻译知识　词义的引申

科技英语的理论准确,所下的定义、定律和定理精确,所描绘的概念、叙述的生产工艺过程清楚。但是在英译汉时,经常会出现某些词在字典上找不到适当的词义的情况。如果生搬硬套,译文则生硬晦涩,不能确切表达原意,甚至有时造成误译。这时就要结合上下文,根据逻辑关系,进行词义引申,才能恰如其分地表达出原意。

1. 概括化或抽象化引申

科技英语常常使用表示具体形象的词来表示抽象的意义。翻译这类词时,一般可将其词义做概括化或抽象化的引申,译文才符合汉语习惯,流畅、自然。例如:

（1）The plan for launching the man-made satellite still lies on the table.

那项发射人造卫星的计划仍被搁置,无法执行。

on the table 按字面意思译成"放在桌子上"语义不通,根据上文意思抽象引申为"无法执行",符合原意。

（2）Military strategy may bear some similarity to the chessboard but it is dangerous to carry the analogy too far.

打仗的策略同下棋可能有某些相似之处,但是如果把这两者之间的类比搞过了头则是危险的。

chessboard 是"棋盘"的意思。棋盘是实物,打仗的策略是思想,不好类比。因此,这里把具体的"棋盘"引申为概括性的"下棋",就说得通了。

（3）The book is too high-powered for technician in general.

这本书对一般技术人员来说也许内容太深。

high-powered 本意为"马力大",引申为"(艰)深"。

（4）The expense of such an instrument has discouraged its use.

这种仪器很昂贵,使其应用受到了限制。

expense 原意为花费,开支,引申为"(仪器)昂贵"。

（5）Industrialization and environmental degradation seem to go hand in hand.

工业化发展似乎伴随着环境的退化。

hand in hand 原意为"携手",引申为"伴随"。

2. 具体化或形象化引申

科技英语中有用代表抽象概念或属性的词来表示一种具体事物。如按字面译,则难以准确表达原文意思。这时就要根据上下文,对词义加以引申,用具体或形象化的词语表达。例如:

（1）Along the equator it reaches nearly halfway around the globe.

它沿着赤道几乎绕地球半周。

reach halfway 意为"达到一半路程"。本句讨论的是围绕地球旋转,根据这一具体语境,可以将 reach halfway 本来含义形象化引申,译为"绕地球半周"。

（2）The shortest distance between raw material and a finished part is precision casting.

把原料加工成成品的最简便的方法是精密铸造。

shortest distance 原意为"最短距离",直接按照这一字面意思,句子有失通顺。可以形象化引申为"最简便的方法"。

(3) The foresight and coverage shown by the inventor of the process are most commendable.

这种方法的发明者所表现的远见卓识和渊博知识,给人以良好的印象。

coverage 原意为"覆盖",引申为"渊博知识"。

(4) The purpose of a driller is to cut holes.

钻床的功能是钻孔。

purpose 原意为"目的",引申为"功能"。

(5) There are many things that should be considered in determining cutting speed.

在测定切削速度时,应当考虑许多因素。

things 原意为"事情",具体引申为"因素"。

Reading Material

阅读下列文章。

Text	Note
OrCAD View **Full-Featured Schematic Editor** OrCAD Capture, a flat[1] and hierarchical Schematic Page Editor, is based on OrCAD's legacy of fast, intuitive[2] schematic editing. Schematic Page Editor combines a standard Windows user interface with functionality and features specific to the design engineer for accomplishing design tasks and publishing design data. • Undo and redo schematic edit unlimited times. • Use Label State for "what if" scenarios[3]. • Launch Property Spreadsheet Editor at design or schematic level to edit or print your design properties. • View and edit multiple schematic designs in a single session. • Reuse design data by copying and pasting within or between schematics. • Select parts from a comprehensive[4] set of functional part libraries. • In-line editing of parts to allow pin name and number movement. • File locking in case the design is being open by another user. **OrCAD Capture** OrCAD Capture® design entry is the most widely used schematic entry system in electronic design today for one simple reason: fast and universal design entry. Whether you're designing a new analog circuit, revising schematic diagram for an existing PCB[5], or designing a digital block diagram with an HDL module, OrCAD Capture provides simple schematic commands you need to enter, modify and verify the design for PCB. • Place, move, drag, rotate, or mirror individual parts or grouped selections while preserving both visual and electrical connectivity[6].	[1] *adj.* 平面的 [2] *adj.* 直觉的 [3] *n.* 情况 [4] *adj.* 全面的,广泛的 [5] Printed Circuit Board, 印制电路板 [6] *n.* 连通性

- Ensure design integrity through configurable Design and Electrical Rule checkers.
- Create custom title blocks and drawing borders to meet the most exacting specifications.
- Insert drawing objects, bookmarks, logos[7] and bitmapped pictures.
- Choose from metric or imperial[8] unit grid spacing to meet all drawing standards.
- Design digital circuits with VHDL or Verilog Text Editor.

Find and select parts or nets quickly from the OrCAD Capture Project Manager and the multi-window interface makes navigation[9] across hierarchy a breeze.

Project Manager Coordinates Design Data

The sophisticated[10] Project Manager simplifies organizing and tracking the various types of data generated in the design process.

An expanding-tree diagram makes it easy to structure and navigate all of your design files, including those generated by PSpice® simulators, Capture CIS and other plug-ins.

- Project Creation Wizard guides you through all the resources available for a specific design flow.
- Centralized management of all design data permits a seamless[11] interchange of schematic data for OrCAD plug-ins and downstream flow.
- Hierarchy browser lets you navigate the entire schematic structure and open specific elements whether it's a schematic page, a part, or net—instantly.
- File tab groups multi-page schematics in folders for flat designs and creates new folders automatically for added levels of hierarchical[12] designs.
- Archive capability ensures the portability[13] of your entire design project.

Hierarchical Design and Reuse

OrCAD Capture boosts[14] schematic editing efficiency by enabling you to reuse subcircuits[15] without having to make multiple copies. Instead, using hierarchical blocks, you can simply reference the same subcircuit multiple times.

- Enables a single instance of the circuitry[16] for you to create, duplicate and maintain.
- Automatic creation of hierarchical ports eliminates potential design connection errors.
- Update ports and pins dynamically for hierarchical blocks and underlying schematics.
- Reuse OrCAD Layout and Cadence® Allegro® high-speed PCB modules within or between schematics.
- Requires just one instance of the circuitry for you to create and maintain.
- Unlimited referencing and reuse of circuitry throughout your entire design.
- Serve schematic pages from library files.
- Sophisticated Property Editor clearly distinguishes[17] properties in a subcircuit from those in referenced uses allowing you to view and edit from one place.

[7] n. 标识
[8] adj. 英制的(度量衡)
[9] n. 向导,导航
[10] adj. 老练的,有经验的
[11] adj. 无缝的,无痕的
[12] adj. 分等级的
[13] n. 可携带,轻便
[14] vt. & vi. 推进
[15] n. 支电路
[16] n. 电路,线路
[17] vt. & vi. 区别,辨别

Libraries And Part Editor

You can access Library Editor directly from the OrCAD Capture user interface. Create and edit parts in the library or directly from the schematic page without interrupting your workflow.

- Movable pin name and pin number.
- Intuitive graphical controls speed of schematic part creation and editing.
- Create new parts quickly by modifying existing ones.
- Spreadsheet[18] and pin array utilities make short work of creating and editing pin-intensive devices.
- Bused vector pins reduce clutter on schematics.
- Create FPGA and CPLD symbols quickly and easily with Part Generator. Compatible with ten popular places and route pin reports.
- Drag-and-drop parts between libraries.
- Speed creation and maintenance of master library sets with design cache.
- Revise a part in the original subcircuit only, or propagate[19] the change to all other uses of the subcircuit in the design.
- Capability to add or delete sections of multisection[20] homogeneous/heterogeneous parts.
- Control power and ground pin visibility and connectivity on a per-schematic basis.

Integrate Huge I/O Count FPGA And CPLD

OrCAD Capture provides a Library Part Generator to automate the integration of FPGA and PLD[21] devices into your system schematic. The Generate Part feature simplifies the creating of core FPGA library parts for devices that might have many hundreds of pins. Signal placement reports created by popular FPGA design applications like those from Altera, Actel, and Xilinx — are read into Generate Part to design the core Capture symbol saving up the hours of tedious graphical entry work. OrCAD Capture supports Xilinx 4.1i/4.2iPAD file format. If, during the PCB layout phase, the PCB designer discovers a more efficient pin placement scheme for the package or additional functionality[22] is added to the FPGA or PLD — the system engineer must modify the symbol and schematics to reflect this change which is error prone and may cause designs to be out of sync. The Generate Part feature has an annotate[23] option which modifies an existing symbol with new pin assignments.

Step 1 Creating parts with potentially hundreds of pins is an error prone and painstaking task. With Generate Part you simply browse in the pin and signal report file created by your place and route software.

Step 2 Specify to create a new part or update an existing one. Packages of all kinds are supported including PGAs and BGAs.

Step 3 The new part is created fast. Pins with common names are intelligently[24] grouped and ordered.

Easy Entry Of Part, Pin, And Net Data

Access all part, net, pin, and title block properties, or any subset, and make changes quickly through the OrCAD Capture spreadsheet Property Editor.

[18] n. 电子表格,电子制表软件,电子数据表

[19] vt. & vi. 繁殖,传播

[20] 多节,多段

[21] [计]可编程逻辑电路

[22] n. 功能性,泛函性

[23] vt. & vi. 注释,评注

[24] adv. 聪明地,智能地

- Select a circuit element, grouped area, or entire page then add/edit/delete part, net, or pin properties.
- Globally apply specific property names across all your designs to meet your particular netlist[25] or other output requirements. This maintains consistency, reduces manual errors, and eliminates multiple re-entry.
- Browse and instantly visit any part, net, hierarchical port, off page connector, bookmark, or design rule error marker from a single reference point.

Verify Circuits Early With Design Rule Check

The configurable Design Rule Check (DRC) feature in OrCAD Capture allows a comprehensive verification of your design before committing[26] to downstream design processes saving the time and cost of ECOs latter in the design cycle.
- Report duplicate parts.
- Identify invalid design packaging.
- Detect off-grid objects leading to unconnected signals.
- Configure with electrical violations to report and assign severity[27] warnings.
- Check entire design or specific modules.

Reports

OrCAD Capture creates basic bill of materials (BOMs) outputs extracting from the information contained in the schematic database.
- Extract all part properties in the schematic design and output them to a text file.
- Automatically package parts with reference designators prior to report generation.

Part Selection

While placing a component, you can identify it visually, modify the properties as needed, then dynamically place it within a design—all in the same sequence.
- Zero-in quickly on the exact library part you want, using wildcard[28] searches.
- Pick your recent part choices from the most recently used (MRU) menu.
- Choose a logic gate or DeMorgan equivalent.
- Edit schematic parts graphically prior to placement.
- Add, modify, and delete part properties at any time.
- Place previously used parts fast by grabbing them from the project design cache.
- Automatically assign reference designators during or after part placement. Update all, or just unidentified[29] parts, or reset all to placeholder values.
- Add libraries to a project from any drive or directory without leaving the part selector.
- Apart filter[30] can be used to filter out the parts from existing libraries based on parameters like HDL models, Spice models, etc. associated with symbols.

[25] n. 连线表

[26] vt. 把……交托给,提交

[27] n. 严格,严重

[28] n. 通配符

[29] adj. 未经确认的

[30] n. 滤波器

Interface Capabilities

OrCAD Capture interfaces with other CAD applications with minimal[31] translation needs or integration problems by importing and exporting virtually every commonly used design file format.

- Export of DXF files to AutoCAD™.
- View and redline schematic with MYRIAD™.
- Bi-directional EDIF 200 graphic transfer and export of the EDIF 200 netlist format.
- Import MicroSim® schematic.
- Export of more than 30 netlist formats, including VHDL, Verilog®, PSpice, SPICE, and PADS 4.0.
- Interface with OrCAD Layout and Allegro PCB with forward and back annotation.
- Interface with NC VHDL Desktop and Synplicity Synplify® for FPGA design.
- Interface with NC VHDL Desktop and NC Verilog® Desktop for board level (multi-chips) digital simulation[32].
- Creation of custom netlists using Microsoft Visual Basic.

Printing And Plotting

Produce professional hardcopy through any output device supported by Microsoft Windows.

- Print Area prints specific area of the design in larger scale.
- Print Preview ensures proper scale and orientation[33].
- Export to the DXF format for CAD interchange.

Cross-probing[34] between OrCAD Capture and Cadence Allegro PCB Layout.

Getting Help

When you have a question or need to accomplish a specific task, the information you need is always a few mouse clicks away.

- Tap into the knowledge of the OrCAD Capture online community.
- Find the answer you need by searching the online help system, and navigate quickly between related topics with extensive hypertext[35] cross references.
- Get up to speed quickly with the award-winning "Learning Capture" online interactive tutorial.
- Jump directly to any topic of interest in the online OrCAD Capture User's Guide in Adobe® Acrobat® format (Adobe Acrobat Reader included).

System Requirements

- Pentium® II 300MHz PC (or faster).
- Windows® XP Professional, Windows® XP Home Edition, Windows® 2000 (SP 2 or higher), or Windows® NT 4.0 (SP6A or higher).
- Minimum 64MB RAM.
- 256MB swap[36] space.
- 256-color Windows® display driver with a minimum of 800×600 resolution (1024×768 recommended).

---

[31] adj. 最小的,最小限度的

[32] n. 仿真,模拟

[33] n. 方向,方位,定位

[34] n. 探测,探查

[35] n. 超文本

[36] n. 交换

参考译文 简单电路

1. 一个电路

电流、电压和电阻之间存在最基本的关系。一个简单的电路包括电源、某些类型的负载和一条让电子在电源和负载之间流动的导线。在下面的电路中,电池提供电压源、电线用作导体、灯泡提供电阻。开关作为电路的附加元件,由此形成一个完整的电路。如果开关断开,电路不通,电灯将不亮。闭合开关,电路接通,则电子通过灯泡从负极流向正极。

(图略)

2. 一个电路示意图

下面是一个电路示意图,包括一个电池、一个电阻、一个电压表和一个电流表。电流表串联在电路中,用于显示有多大的电流流过电路。电压表跨接于电压源,用于显示电池提供多大的电压。在分析一个电路的组成之前,我们先了解欧姆定律。

(图略)

3. 欧姆定律

19 世纪德国数学家乔治·西蒙·欧姆研究了电流、电压和电阻之间的关系,他用公式表示成一条定律来说明三者间的关系:电流同电压变化成正比,同电阻变化成反比。通过该定律可以导出下面的公式。

(公式略)

欧姆定律是适用于全部电路的基本定律。电气设计者必须决定对于给定的负载需要多大的电压,如计算机、时钟、灯和电动机。这种决定必定涉及电流、电压和电阻之间的关系。所有的电气设计和分析都是由欧姆定律开始的。有以下三种表达欧姆定律的数学方法。

(公式略)

使用哪个公式决定于开始知道什么,以及需要知道什么。

4. 欧姆定律三角形

有一种简单的方法来记忆使用哪个公式。通过将电流、电压和电阻安排在一个三角形中,就可以迅速确定正确的公式。

(图略)

5. 利用三角形

要利用这个三角形,盖住你想要计算的值。用剩下的字符组成公式。

(图略)

只有在使用正确的值时欧姆定律才能给出正确的答案。记住下面的三个规则:
- 电流总是用安培表示的。

- 电压总是用伏特表示的。
- 电阻总是用欧姆表示的。

直流串联电路

1. 在串联电路中的电阻

当任意个电阻头尾相连构成只有一条电流能流过的路径时，就形成串联电路。电阻可以是实际的电阻器或者是有电阻的其他设备。如图（图略）所示为4个电阻头尾连接。只有一条电流经过的路径：从负极通过 R_4、R_3、R_2 和 R_1 返回到正极。

（图略）

2. 串联电阻公式

在串联电路中电阻阻值是相加的。如果一个 4Ω 的电阻器同一个 6Ω 的电阻器连在一起，总电阻值是 10Ω。别的有阻抗的设备串联也是这样的。电阻串联的数学公式是

（公式略）

给出的串联电路中，R_1 是 $11k\Omega$，R_2 是 $2k\Omega$，R_3 是 $2k\Omega$，R_4 是 100Ω，R_5 是 $1k\Omega$，电路的总电阻是多少？

（计算过程略）

3. 串联电路的电流

串联电路中总电阻的公式使我们能够简化电路。利用欧姆定律就可以计算出电流的值。在串联电路中的任何地方测量电流的值都是相同的。

（公式及图略）

4. 串联电路中的电压

电压可以通过电路中每个电阻上的压降来测量。电压经过一个电阻器就意味着一个电压降。德国物理学家基尔霍夫阐述了一个定律：整个回路中各个电阻器上的电压降的总和等于给这个回路提供的电压。在下面的示意图中，4个阻值都为 1.5Ω 的电阻串联起来连接一个 $12V$ 的电池。根据欧姆定律可以知道每个电阻器上"下降"相等的电压。

（图略）

第一，求解总电阻。
（公式及计算略）
第二，求解电流。
（公式及计算略）
第三，求解任意电阻上的电压。
（公式及计算略）

测量任意一个电阻器，电压表的读数都会是 $3V$。如果测量 R_3 和 R_4 组合上的电压，电压表的读数是 $6V$。如果测量 R_2、R_3 和 R_4 组合上的电压，电压表的读数是 $9V$。如果将4个电阻器上的电压降加起来，其和将是给电路提供电压的电池的电压，即 $12V$。

5. 串联电路上电压的划分

人们往往希望使用低于供电电压的电压。为此,可以使用电压分配器。电池电压用 U_1 表示,本例中为50V。希望得到的电压用 U_0 表示。通过计算得出是40V。为了计算这个电压,第一求解总电阻。

(公式及计算略)

第二,求解电流。

(公式及计算略)

第三,求解电压。

(公式及计算略)

Unit 4

Text A Basic Semiconductor Crystal Structure

To understand how diodes, transistors, and other semiconductor devices can do what they do, it is first necessary to understand the basic structure of all semiconductor devices. Early semiconductors were fabricated from the element germanium, but silicon is preferred in most modern applications.

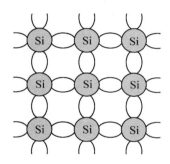

The crystal structure of pure silicon is of course 3-dimensional, but that is difficult to display or to see, so the image to the left is often used to represent the crystal structure of silicon. For you physics types, silicon (and germanium) falls in column Ⅳa of the Periodic Table. This is the carbon family of elements. The essential characteristic of these elements is that each atom has four electrons to share with adjacent atoms in forming bonds.

While this is an oversimplified description, the nature of a bond between two silicon atoms is such that each atom provides one electron to share with the other.[1] The two electrons thus shared are in fact shared equally between the two atoms. This type of sharing is known as a covalent bond. Such a bond is very stable, and holds the two atoms together very tightly, so that it requires a lot of energy to break this bond.

For those who are interested, the actual bonds in a 3-dimensional silicon crystal are arranged at equal angles from each other. If you visualize a tetrahedron (a pyramid with three points on the ground and a fourth point sticking straight up) with the atom centered inside, the four bonds will be directed towards the points of the tetrahedron.

Now we have known our silicon crystal, but we still haven't known semiconductor. In the figure we saw above, all of the outer electrons of all silicon atoms are used to make covalent bonds with other atoms. There are no electrons available to move from place to place as an electric current. Thus, a pure silicon crystal is quite good insulator. In fact, it is almost glass, which is silicon dioxide. A crystal of pure silicon is said to be an intrinsic crystal.

To allow our silicon crystal to conduct electricity, we must find a way to allow some electrons

to move from place to place within the crystal, in spite of the covalent bonds between atoms. [2]
One way to accomplish this is to introduce an impurity such as arsenic or phosphorus into the crystal structure, as shown to the right. These elements are from column Va of the Periodic Table, and have five outer electrons to share with other atoms. In this application, four of these five electrons bond with adjacent silicon atoms as before, but the fifth electron cannot form a bond. This electron can easily be moved with only a small applied electrical voltage. Because the resulting crystal has an excess of current-carrying electrons, each with a negative charge, it is known as "N-type" silicon. [3]

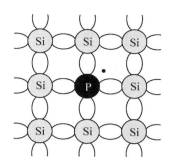

This construction does not conduct electricity as easily and readily as, say, copper or silver. It does exhibit some resistance to the flow of electricity. It cannot properly be called a conductor, but at the same time it is no longer an insulator. Therefore, it is known as a semiconductor.

While this effect is interesting, it still isn't particularly useful by itself. A plain carbon resistor is easier and cheaper to manufacture than a silicon semiconductor. We still don't have any way to actually control electrical current.

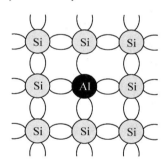

But wait a moment! We obtained a semiconductor material by introducing a 5-electron impurity into a matrix of 4-electron atoms. (For you physics types, we're only looking at the outer electrons that are available for bonding—electrons in inner shells are not included in the process or in this discussion.) What happens if we go the other way, and introduce a 3-electron impurity into such a crystal? Suppose we introduce some aluminum (from column Ⅲa in the Periodic Table) into the crystal, as shown to the left. We could also try gallium, which is also in column Ⅲa right under aluminum. Now what?

These elements only have three electrons available to share with other atoms. Those three electrons do indeed form covalent bonds with adjacent silicon atoms, but the expected fourth bond cannot be formed. A complete connection is impossible here, leaving a "hole" in the structure of the crystal.

Experimentation shows that if there is an empty place where an electron should logically go, often an electron will try to move into that space to fill it. However, the electron filling the hole had to leave a covalent bond behind to fill this empty space, and therefore leaves another hole behind as it moves. [4] Yet another electron may move into that hole, leaving another hole behind, and so forth. In this manner, holes appear to move as positive charges through the crystal. Therefore, this type of semiconductor material is designated as "P-type" silicon.

By themselves, P-type semiconductors are no more useful than N-type semiconductors. The truly interesting effects begin when the two are combined in various ways, in a single crystal of silicon. The most basic and obvious combination is a single crystal with an N-type region at one end and a P-type region at the other. A crystal with two regions as described is known as a semiconductor diode.

New Words and Phrases

diode	['daiəud]	n.	二极管
transistor	[træn'sistə]	n.	晶体管
semiconductor	['semikən'dʌktə]	n.	半导体
fabricate	['fæbrikeit]	vt.	制作,构成
germanium	[dʒəː'meiniəm]	n.	锗
physics types			物理类型,物理结构
silicon	['silikən]	n.	硅,硅元素
fall in			属于
carbon family			碳族
oversimplify	['əuvə'simplifai]	vt. & vi.	(使)过分地单纯化
dimensional	[di'menʃənəl]	adj.	空间的
adjacent	[ə'dʒeisənt]	adj.	邻近的,接近的
bond	[bɔnd]	n.	结合(物),粘结(剂)
		vt. & vi.	结合
covalent	[kəu'veilənt]	adj.	共有原子价的,共价的
3-dimensional			三维的
tetrahedron	['tetrə'hedrən]	n.	四面体
pyramid	['pirəmid]	n.	角锥,棱锥
crystal	['kristl]	n.	晶体,水晶,结晶
		adj.	结晶状的
outer electron			外层电子
intrinsic	[in'trinsik]	adj.	(指价值、性质)固有的,内在的,本质的
in spite of		adv.	不管
impurity	[im'pjuəriti]	n.	杂质,混杂物,不洁,不纯
N-type			N 型
arsenic	['ɑːsənik]	n.	[化]砷,砒素
phosphorus	['fɔsfərəs]	n.	磷
periodic Table			(元素)周期表
experimentation	[eks,perimen'teiʃən]	n.	实验,实验法
gallium	['gæliəm]	n.	镓
positive charge			正电荷
P-type			P 型

Notes

[1] While this is an oversimplified description, the nature of a bond between two silicon atoms is such that each atom provides one electron to share with the other.

本句中,while 有"尽管"的意思。例如,While I understand what you say, I can't agree with you. 虽然我理解你的意思,但我还是不同意。nature 意为"性质,属性"。nature of a bond between silicon atoms 指两个硅原子结合的性质。

[2] To allow our silicon crystal to conduct electricity, we must find a way to allow some electrons to move from place to place within the crystal, in spite of the covalent bonds between atoms.

本句中,to allow our silicon crystal to conduct electricity 是动词不定式短语作目的状语,to allow some electrons to move from place to place within the crystal 作定语,修饰 a way, in spite of 表示"尽管有……"。

[3] Because the resulting crystal has an excess of current-carrying electrons, each with a negative charge, it is known as "N-type" silicon.

本句中的 resulting 是"作为结果的,因而发生"的意思。因此理解这句话需要与上句结合起来,因为这句话中的晶体是上句话变化的结果。each with a negative charge 是伴随状况,表明每个晶体都有负电荷。由于以上的原因,它就被称为"N 型"硅。

[4] However, the electron filling the hole had to leave a covalent bond behind to fill this empty space, and therefore leaves another hole behind as it moves.

本句中,however 表示转折,主语是 the electron,谓语是 had to leave,宾语是 a covalent bond,to fill this empty space 为目的状语,filling the hole 修饰的是 the electron。therefore 表示结果,as 引导了一个时间状语从句。

Exercises

【Ex.1】 根据课文内容,回答以下问题。

(1) What element were early semiconductors fabricated from?

(2) How are the actual bonds in a 3-dimensional silicon crystal arranged?

(3) Which column is arsenic and phosphorus from in the Periodic Table, and how many outer electrons do they have to share with other atoms?

(4) What's the feature of the elements that are from column Ⅲa in the Periodic Table?

(5) What's the structural feature of the "P-type" silicon?

【Ex. 2】 根据下面的英文解释，写出相应的英文词汇。

英 文 解 释	词　汇
an electronic device that restricts current flow chiefly to one direction	
a small electronic device containing a semiconductor and having at least three electrical contacts, used in a circuit as an amplifier, a detector, or a switch	
any of various solid crystalline substances, such as germanium or silicon, having electrical conductivity greater than insulators but less than good conductors	
a homogenous solid formed by a repeating, 3 dimensional pattern of atoms, ions, or molecules and having fixed distances between constituent parts	
a tabular arrangement of the elements according to their atomic numbers so that elements with similar properties are in the same column	
something, such as a fetter, cord, or band, that binds, ties, or fastens things together	
a material that insulates, especially a nonconductor of sound, heat, or electricity	
a rectangular array of numeric or algebraic quantities subject to mathematical operations	
a compound with two oxygen atoms per molecule	
the quality or condition of being impure, especially	

【Ex. 3】 把下列句子翻译为中文。

(1) Generally transistors fall into the category of bipolar transistor, either the more common NPN bipolar transistors or the less common PNP transistor types.

(2) Silicon crystals for example have very few free electrons. However if "impurities" (different atomic structure—e. g. arsenic) are introduced in a controlled manner then the free electrons or conductivity is increased.

(3) If we take a piece of the P-type material and connect it to a piece of N-type material and apply voltage then current will flow. Electrons will be attracted across the junction of the P and N materials.

(4) A silicon flake a quarter inch on a side can hold a million electronic components, ten times more than 30 ton ENIAC, the world's first electronic digital computer.

(5) A microprocessor, for example, can endow a machine with decision-making ability, memory for instructions, and self-adjusting controls.

(6) Eventually one billion transistors, or electronic switches, many crowd into a single chip. A memory chip of such complexity could store the text of 200 long novels.

(7) A transistor can be configured as a diode and often are used in certain projects, especially to adjust for thermal variations.

(8) A transistor is often a limitation and at other times an asset that with zero spacing between the P and N junctions we have a relatively high value capacitor.

(9) The early principal application of diodes was in rectifying 50/60Hz AC mains to raw DC which was later smoothed by choke transformers or capacitors.

(10) Varactor or tuning diodes work on the principle that all diodes exhibit some capacitance.

【Ex.4】 把下列短文翻译成中文。
The transistor was developed at Bell Laboratories in 1948. Large scale commercial use didn't come until much later owing to slow development. Transistors used in most early entertainment equipments were the germanium types. When the silicon transistor was developed it took off dramatically. The first advantages of the transistor were relatively low power consumption at low voltage levels which made large scale production of portable entertainment devices feasible. Interestingly the growth of the battery industry has paralleled the growth of the transistor industry.

【Ex.5】 通过Internet查找资料,借助如"金山词霸"等电子词典和辅助翻译软件,完成以下技术报告。通过E-mail发送给老师,并附上你收集资料的网址。
(1) 当前世界上有哪些最主要的半导体生产厂家以及有哪些最新型的产品(附上各种最新产品的图片)。

（2）当前世界上有哪些最主要的计算机芯片生产厂家以及有哪些最新型的产品（附上各种最新产品的图片）。

Text B The PN Junction

We've seen that it is possible to turn a crystal of pure silicon into a moderately good electrical conductor by adding an impurity such as arsenic or phosphorus (for an N-type semiconductor) or aluminum or gallium (for a P-type semiconductor). By itself, however, a single type of semiconductor material isn't very useful. Useful applications start to happen only when a single semiconductor crystal contains both P-type and N-type regions. Here we will examine the properties of a single silicon crystal which is half N-type and half P-type.

Consider the silicon crystal represented to the right. Half is N-type while the other half is P-type. We've shown the two types separated slightly, as if they were two separate crystals. The free electrons in the N-type crystal are represented by small black circles with a " – " sign inside to indicate their polarity. The holes in the P-type crystal are shown as small white circles with a " + " inside.

In the real world, it isn't possible to join two such crystals together usefully. Therefore, a practical PN junction can only be created by inserting different impurities into different parts of a single crystal. So let's see what happens when we join the N-type and P-type crystals together, so that the result is one crystal with a sharp boundary between the two types.

You might think that, left to itself, it would just sit there. However, this is not the case. Instead, an interesting interaction occurs at the junction. The extra electrons in the N region will seek to lose energy by filling the holes in the P region. This  leaves an empty zone, or depletion region as it is called, around the junction as shown to the right. This action also leaves a small electrical imbalance inside the crystal. The N region loses some electrons so it has a positive charge. Those electrons have migrated to fill holes in the P region, which therefore has a negative charge. This electrical imbalance amounts to about 0.3V in a germanium crystal, and about 0.65 to 0.7V in a silicon crystal. This will vary somewhat depending on the concentration of the impurities on either side of the junction.

Unfortunately, it is not possible to exploit this electrical imbalance as a power source; it doesn't work that way. However, we can apply an external voltage to the crystal and see what happens in response. Let's take a look at the possibilities.

Suppose we apply a voltage to the outside ends of our PN crystal. We have two choices. In  this case, the positive voltage is applied to the N-type material. In response, we see that the positive voltage applied to the N-type material attracts any free electrons towards the end of the crystal and away from the junction, while the negative voltage applied to the P-type end attracts holes away from the junction on this end. The result is that all available current carriers are attracted away from the junction, and the depletion region grows correspondingly larger. There is no current flow through the crystal because all available current carriers are attracted away from the junction, and cannot cross. (We are here considering an ideal crystal—in real life, the crystal can't be perfect, and some leakage current does flow.) This is known as reverse bias applied to the semiconductor crystal.

Here the applied voltage polarities have been reversed. Now, the negative voltage applied to the N-type end pushes electrons towards the junction, while the positive voltage at the P-type end pushes holes towards the junction. This has the effect of shrinking the depletion region. As the applied voltage exceeds the internal electrical imbalance, current carriers of both types can cross the junction into the opposite ends of the crystal.
Now, electrons in the P-type end are attracted to the positive applied voltage, while holes in the N-type end are attracted to the negative applied voltage. This is the situation of forward bias.

Because of this behavior, an electric current can flow through the junction in the forward direction, but not in the reverse direction. This is the basic nature of an ordinary semiconductor diode.

It is important to realize that holes exist only within the crystal. A hole reaching the negative terminal of the crystal is filled by an electron from the power source and simply disappears. At the positive terminal, the power supply attracts an electron out of the crystal, leaving a hole behind to move through the crystal toward the junction again.

In some literature, you might see the N-type connection designated as the cathode of the diode, while the P-type connection is called the anode. These designations come from the days of vacuum tubes, but are still in use. Electrons always move from cathode to anode inside the diode.

One point that needs to be recognized is that there is a limit to the magnitude of the reverse voltage that can be applied to any PN junction. As the applied reverse voltage increases, the depletion region continues to expand. If either end of the depletion region approaches its electrical contact too closely, the applied voltage has become high enough to generate an electrical arc straight through the crystal. This will destroy the diode.

It is also possible to allow too much current to flow through the diode in the forward direction. The crystal is not a perfect conductor, remember; it does exhibit some resistance. Heavy current flow will generate some heat within that resistance. If the resulting temperature gets too high, the semiconductor crystal will actually melt, again destroying its usefulness.

Always be careful to pay attention to the maximum specifications of a diode, and be sure to keep the operating conditions of the diode well within the indicated limits.

New Words and Phrases

junction ['dʒʌŋkʃən] n. 连接,接合,交叉点,汇合处

PN junction		PN 结
moderate	['mɔdərit]	*adj.* 中等的,适度的,适中的
		vt. & vi. 缓和
boundary	['baundəri]	*n.* 边界,分界线
interaction	[,intər'ækʃən]	*n.* 交互作用,交感
depletion	[di'pli:ʃən]	*n.* 损耗
imbalance	[im'bæləns]	*n.* 不平衡,不均衡
N region		N 区
P region		P 区
migrate	[mai'greit, 'maigreit]	*vi.* 迁移,移动,移往,移植
		vt. 使移居,使移植
concentration	[,kɔnsen'treiʃən]	*n.* 集中,集合,专心
exploit	[iks'plɔit]	*vt.* 开拓,开发,开采
response	[ris'pɔns]	*n.* 回答,响应,反应
reverse	[ri'və:s]	*n.* 相反,背面,反面,倒退
		adj. 相反的,倒转的,颠倒的
		vt. 颠倒,倒转
bias	['baiəs]	*n.* 偏见,偏爱,斜线
		vt. 使存偏见
vacuum	['vækjuəm]	*n.* 真空,空间,真空吸尘器
		adj. 真空的,产生真空的,利用真空的
leakage	['li:kidʒ]	*n.* 漏,泄漏,渗漏

Exercises

【Ex. 6】 根据文章所提供的信息判断正误。

(1) We introduce some aluminum (from column Ⅲa in the Periodic Table) into the crystal, it will be turned into a moderately good electrical conductor.

(2) A single type of semiconductor material is very useful.

(3) A practical PN junction can only be created by inserting any impurities into different parts of a single crystal.

(4) When we join the N-type and P-type crystals together, they will produce one crystal with a sharp boundary between the two types.

(5) When we join the N-type and P-type crystals together, the extra electrons in the P region will seek to lose energy by filling the holes in the N region.

(6) In a PN crystal, the N region loses some electrons so it has a positive charge and those electrons have migrated to fill holes in the P region, which therefore has a negative charge.

(7) We can exploit this electrical imbalance as a power source.

(8) Suppose we apply a voltage to the outside ends of a PN crystal, if the positive voltage is applied to the N-type material and the negative voltage applied to the P-type, there is no current

flow through the crystal.

(9) An electric current can flow through the junction in the forward direction, or in the reverse direction.

(10) If we apply a high reverse voltage to a diode, it will be destroyed.

科技英语翻译知识　词义的增减

英汉两种语言,有不同的表达方式,在翻译过程中,要对语义进行必要的增减。当句子结构不完善,句子含义不明确或词汇概念不清晰时,需要对语意加以补足。反之,原文中的有些词如果在译文中不言而喻,就要省略一些不必要的词,使得译文更加严谨、明确。

1. 增补

增补的基本功能是明确原文词汇含义,表达原文语法概念,满足译文修辞要求。

(1) The dielectric material in a practical capacitor is not perfect, and a small leakage current will flow through it.

实际应用的电容电介材料是不完全绝缘的,因而会通过小的漏电流。

not perfect 的意思是"不完全的、不完美的",但在哪方面不完全,没有给出,因此要补足原文词汇含义"绝缘"。

(2) Electricity is convenient and efficient.

电用起来方便而有效。

增加"用起来",明确原文含义。

(3) The high-altitude plane was and still is a remarkable bird.

高空飞机过去是而且现在还是一种了不起的飞行器。

增加"过去"、"现在"二词,表达动词过去时和现在时的语法概念。

(4) It is estimated that the new synergy between computers and Net technology will have significant influence on the industry of the future.

有人预测,新的计算机和网络技术的结合将会对未来工业产生巨大的影响。

增加主语"有人",使句子变成主动句。

(5) The solution of parallel AC circuits problems is different from series AC circuits.

并联交流电路问题的解答方法,不同于串联交流电路。

汉语不习惯用动词作主语,因此补充"……的方法",满足修辞要求。

(6) This magnetic field may be that of a bar magnet, an U-shaped magnet, or an electromagnet.

这个磁场可以是条形磁铁的磁场,可以是马蹄形磁铁的磁场,也可以是电磁铁的磁场。

译文两次重复"可以是"和"磁场",以强调有三种可能性。

(7) Now human beings have not yet progressed as to be able to make an element by combining protons, neutrons and electrons.

目前人类没有进展到能把质子、中子、电子三者化合成为一个元素的地步。

增加概括性数量词"三者"。

（8）Using a transformer, power at low voltage can be transformed into power at high voltage.

如果使用变压器，低电压的电力就能转换成高电压的电力。

增加"如果"，使语气连贯。

（9）Were there no electric pressure in a conductor, the electron flow would not take place in it.

导体内如果没有电压，便不会产生电子流动现象。

增加"现象"一词，符合汉语习惯。

2. 省略

英语中的冠词、代词和连词在译文中往往可以省略。

（1）Hence the unit of electric current is the Ampere.

因此，电流的单位是安培。

两个定冠词 the 省略。

（2）The diameter and the length of the wire are not the only factors to influence its resistance.

导线的直径和长度不是影响电阻的唯一因素。

省略物主代词 its。

（3）Like charges repel each other while opposite charges attract.

同性电荷相斥，异性电荷相吸。

省略连词 while。

（4）When atoms are joined together they form a larger particle called a molecule.

原子连接在一起就形成了更大的粒子，叫做分子。

省略连词 when。

（5）When we talk of electric current, we mean electrons in motion.

当我们谈到电流时，我们指的是运动的电子。

省略介词 in。

（6）Evidently semiconductors have a lesser conducting capacity than metals.

显然，半导体的导电能力比金属差。

省略动词 have。

（7）There are many kinds of atoms, differing in both mass and properties.

原子种类很多，质量与性质都不相同。

省略引导词 there。

（8）The invention of radio has made it possible for mankind to communicate with each other over a long distance.

无线电的发明使人类有可能进行远距离通信。

省略形式宾语的引导词 it。

（9）Insulators in reality conduct electricity but, nevertheless, their resistance is very high.

绝缘体实际上也能导电，但其电阻很高。

省略重复性词语 nevertheless。

(10) Semiconductors devices have no filament or heaters and therefore require no heating power or warmed up time.

半导体器件没有灯丝和加热器，因此不需要加热功率或加热时间。

省略重复性词语 heater。

Reading Material

阅读下列文章。

Text	Note
P-CAD 2002 Overview（1） **1. P-CAD 2002 Design environment** 　　P-CAD 2002 gives you full access to the power of the Windows environment. Long file names, tool tips[1], prompt line fly-bys, and context-sensitive help all contribute to a design environment which makes you instantly feel at home. Dockable/floating/resizable toolbars[2], recently used file picks[2], tabbing[3] to status line fields and keyboard shortcuts[4] are just a few of the familiar features that make P-CAD 2002 an environment with productivity[5] firmly in mind. 　　Taking full advantage of the graphical nature of the Windows environment, you can query, move, copy, delete, and modify practically anything on the screen—with full context-sensitivity[6]. Double-click[7] an item to directly view and edit its properties. To accurately choose the objects you want to edit, P-CAD 2002 gives you an extensive range of powerful selection features, including selection-by-query, block selection, and selection of highlighted[8] objects. 　　With P-CAD 2002 you customize[9] the work environment to suit your needs and control design output to meet your standards. Reports can be fully customized to define the fields that are output, under what conditions, and in which order; custom pad designs are supported for those special shapes[10] that you might need; and metric and imperial units are readily mixed to support worldwide operations. Define custom tool bars[11] to allow direct access of external applications from the P-CAD menu[12] and tool bar. User-defined fields and attributes give you full control over the data saved with your design, and actions based on that data. 　　P-CAD 2002's PCB editor supports up to 999 design layers[13], giving you virtually unlimited[14] design potential[15]. You have total flexibility in adding, naming, numbering, and assigning[16] layers to your PCB design. P-CAD 2002 supports layer names with up to 30 characters to make your design and reports easy to view and read, and you can easily enable or disable individual layers for printing and/ or viewing. The ability to fully define the properties of custom layers gives you unbounded[17] flexibility in the definition of your board. For each custom layer you can set its type (signal, non-signal, plane), routing bias[18], color name and number. 　　P-CAD 2002 Design Environment gives you: 　　● Full access to the power of the Windows environment. 　　● Extensive online[19] help and documentation.	[1] *n.* 提示，技巧 [2] *vt. & vi.* 拾取 [3] *n.* 制表，标号 [4] *n.* 快捷键 [5] *n.* 生产力，生产率 [6] *n.* 敏感，灵敏（度），灵敏性 [7] *vt. & vi.* 鼠标双击 [8] *adj.* 高亮的 [9] *vt. & vi.* 定制，用户化 [10] *n.* 外形，形状 [11] *n.* 工具栏 [12] *n.* 菜单 [13] *n.* 层 [14] *adj.* 无限制的，无约束的 [15] *n.* 潜能 [16] *vt.* 分配，指派 [17] *adj.* 极大的 [18] *n.* 偏置，斜线 [19] *adj.* 联机的，在线的

- Up to 999 layers: 11 default[20] and 988 user-defined.
- Up to 30 character net, net class and layer names.
- Support for multiple internal planes: plane layers can be assigned to any net.
- User-definable layer ordering used for DRC, NC drill, Gerber.
- Design rules defined per design, layer, net class, net, class-to-class, room and component.
- Powerful selection features including selection-by-query, block selection, and selection of highlighted objects.
- Fully customizable work environment with user-definable fields[21] and attributes[22].
- Fast application startup.
- Optimized memory management.

2. P-CAD 2002 Schematic capture with power and precision

To stay competitive and at the cutting edge you need the freedom to tackle[23] designs of any complexity.

P-CAD 2002's powerful schematic[24] editor works with you to create even the most detailed hierarchical multi-sheet schematics. Providing support for up to 999 sheets[25], the schematic editor's powerful interactive[26] placement[27] and editing tools give you all the power you need to support your most complex designs.

Make the most of your design time with advanced features such as the ability to easily modify the shape of lines and wires by simply adding vertices, cut-copy-paste of entire circuits, an intuitive select-edit methodology[28] for streamlining design changes, and hot-linked cross-probing with P-CAD PCBs. You can easily open multiple windows to view various schematics simultaneously[29] or have different views of the same schematic.

Context-sensitive, direct access to frequently-used commands, keyboard shortcuts and browse and selection features all work to eliminate costly delays, while comprehensive reporting, output generation and design navigation[30] capabilities mean you can be assured of accurate[31], reliable[32] schematic creation.

P-CAD 2002's schematic editor provides advanced linking between your schematic and PCB layout, allowing you to transfer design data, automatically cross-probe between documents, and forward[33] and back-annotate your design.

Schematic Editor feature highlights:
- Support for multi-level and multi-sheet schematics.
- Up to 999 sheets per design.
- Over 27 000 supplied schematic components with free component library updates[34] available from this web site.
- Hierarchical[35] design rules assigned per design, layer, net class, net, room, component, and class-to-class.
- Advanced linking between schematic and PCB[36] layout.
- User-definable hot keys for all actions.
- Advanced editing capabilities.
- Schematic symbol wizard and library manager.

[20] n. 默认(值)

[21] n. 域
[22] n. 属性

[23] vt. 处理,解决
[24] n. 示意图,原理图
[25] n. 表,单
[26] adj. 交互式的
[27] n. 放置,布置
[28] n. 方法学,方法论
[29] adv. 同时地

[30] n. 导航,向导
[31] adj. 准确的,精确的
[32] adj. 可靠的,可信赖的
[33] vt. & vi. 转发,转递
[34] vt. & vi. 更新,升级
[35] adj. 分等级的
[36] 印制电路板

3. P-CAD 2002 PCB Design Overview

P-CAD 2002 provides a highly versatile[37] and flexible environment for PCB layout and editing. Difficult concepts such as pad stacks and blind/buried vias are addressed with logical, intuitive[38] dialog boxes and menus. Prompt and status lines keep you informed with extensive online information, while the graphical icon-based toolbars provides single-click access to common commands and settings.

With P-CAD 2002's PCB editor, loading a netlist[39] adds connections, components and attributes for nets and components to your PCB design. Missing patterns are no problem. P-CAD 2002 creates a placeholder[40] pattern for any components in the netlist that don't have an attached pattern, allowing it to load without an error. You can then assign correct patterns at a later stage. Such successive disclosure readily supports adding data whenever it is available during the design cycle.

When placing components it is precision and flexibility that counts, and P-CAD 2002 gets the job done with ease. Graphically browse library patterns during placement and quickly change a component's reference designator[41]. You can flip components, move components by designators, rotate entries on user-defined increments to 0.1 degrees, or replace one or more components by type. With P-CAD 2002 the power is in your hands.

Difficult component placement challenges arise from dense[42] PCBs and short design cycles. P-CAD 2002 meets these challenges head-on by providing interactive placement and organization tools that make correct component placement quick and easy.

P-CAD 2002 provides Visual Placement Areas that show clearly where a component can be placed without violating design constraints. Physical, electrical, and manufacturing rules can be analyzed simultaneously or independently.

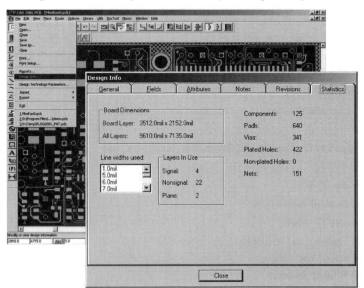

P-CAD 2002 provides a superb PCB editing environment that gives you the power and control necessary to meet today's PCB design challenges.

[37] *adj.* 通用的

[38] *adj.* 直觉的

[39] *n.* 连线表

[40] *n.* 占位符

[41] *n.* 指示器

[42] *adj.* 密集的, 浓厚的

PCB design environment feature highlights：
 • "60×60" maximum board size with 64 000 pad styles per design.
 • User-defined grids available down to 0.0001 inch (0.1 mil) or 0.01 mm (10 micron[43]) resolution：grid settings saved with design.
 • User-definable layer ordering used for Design Rule Checking (DRC), NC drill, Gerber.
 • Support for multiple internal planes：plane layers can be assigned to any net.
 • User-definable board, net and component attributes.
 • Full PCB-Schematic synchronization.
 • Extensive DRC features ensuring board integrity through to manufacture.
 • Fully integrated manual, interactive and automatic routing tools.
 4. P-CAD 2002 Design Verification
     Reducing design iterations[44] saves you time and money. P-CAD 2002 includes extensive DRC features that ensure your board's integrity through to manufacture. The batch DRC checks for clearance, silk-screen, copper pour, drill, plane, text, board edge clearance, and netlist violations[45]. It also checks for unrouted nets, unconnected pins, and "less-than-complete" electrical connections.
     You can check clearances layer-by-layer and item-to-item to ensure your minimum clearances are met in all circumstances.
     P-CAD 2002 includes hierarchical design rules for class-to-class, net, net class, and design level precedence. To find, evaluate and fix errors quickly and efficiently, DRC error indicators track error type and location. Combined with P-CAD 2002's sophisticated[46] filtering and error validation features, these error markers pinpoint trouble spots fast.
     P-CAD 2002 also includes a powerful on-line DRC features that monitors design rules as you move or place items, allowing you to correct errors on-the-fly. Instant violation notification[47] reduces overall design time and re-work.
 Design Verification highlights：
 • Full design rule, DRC and reporting support for test points.
 • Online clearance rules checking during manual and interactive routing.
 • Hierarchical design rules for class-to-class, net, net class, and design level precedence.
 • User-defined pad-to-pad, pad-to-line, pad-to-via, line-to-line, line-to-via, and via-to-via clearances.
 • Layer-by-layer clearance checking.
 • Flag less-than-complete electrical connections.
 • Blind/buried via connectivity[48] check.
 • Check routed trace width against width attribute.
 • Error indicators track error type and location.
 5. P-CAD 2002 Advanced Editing features
     P-CAD 2002's advanced editing features put you in control of the PCB design process. Extensive cut/copy/paste support includes support for copying complete
	[43] *n.* 微米
	
	 [44] *n.* 反复
	
	 [45] *n.* 违规
	
	 [46] *adj.* 改进的，复杂的
	
	 [47] *n.* 通知
	
	 [48] *n.* 连通性

circuits, with control over reference designator and net naming, plus the handling of net attributes to favor the design or circuit. Such expediency[49] for duplicating[50] circuitry is particularly valuable for full or partial design re-use, such as the creation of multiple channels for an audio application.

To change the pattern assigned to a component, P-CAD 2002 allows you to select from multiple pattern graphics pre-assigned to a component, or to browse and select from a list of patterns with matching pad counts. When swapping patterns a new component is created on-the-fly and replaces the original one. The component has the same gate[51] and pin designations as the original, and its attributes can be viewed, changed, deleted, or added before it is saved. In this manner, P-CAD 2002 further supports the philosophy[52] of adding data whenever it becomes available.

For better placement results, connection lines are dynamically optimized when components are placed or moved so you can instantly determine the best component positions.

When reconnecting nets, you can optionally add a system-assigned net name to "free copper" when the copper connects two or more non-net nodes (nodes not already assigned to a net).

[49] n. 方便
[50] n. 复制

[51] n. 逻辑门
[52] n. 基本理论

P-CAD 2002's powerful parametric pattern graphics feature allows you to assign multiple graphics to a single pattern.

ECOs generated when splitting nets will only remove copper tracks local to the change to minimize re-routing.

The features above are just some of the powerful PCB editing features that make P-CAD 2002 the smart choice for PCB professionals[53].

Editing features highlights:

[53] n. 专业人员

- Extensive cut/copy/paste support includes support for copying complete circuits.
- To change the pattern assigned to a component, P-CAD 2002 allows you to select from multiple pattern graphics pre-assigned to a component.
- Parametric pattern graphics allow automatic selection of different graphic based on component orientation[54], layer and solder flow direction.
- Ability to fully lock in place arcs, lines, pads, vias, test points, copper pours and components.
- Reloading netlist adds or deletes components, reconnects existing copper; allows net class and attribute override control.
- Component pads ghosted during placement.
- Hot key to change component reference designator during placement.

6. P-CAD 2002 Powerful Interactive Placement tools with precise component positioning

When placing components it is precision and flexibility that counts, and P-CAD 2002 gets the job done with ease. Graphically browse library patterns during placement and quickly change a component's reference designator. You can flip components, move components by designators, rotate entries on user-defined increments to 0.1°, or replace one or more components by type. With P-CAD 2002 the power is in your hands.

With P-CAD 2002's powerful parametric[55] pattern graphics feature you can assign multiple graphics to a single pattern and define different graphics to be automatically displayed as you rotate a component or swap layers during placement, or according to the specified direction of solder[56] flow. You can also manually[57] swap pattern graphics for a component at any time—very useful for tight spaces when you need to insert a resistor on its end, for example.

Difficult component placement challenges arise from dense PCBs and short design cycles. P-CAD 2002 meets these challenges head-on by providing interactive placement and organization tools that make correct component placement quick and easy. P-CAD 2002 provides Visual Placement Areas that show clearly where a component can be placed without violating design constraints[58]. Physical, electrical, and manufacturing rules can be analyzed simultaneously or independently. For example, the placement area for a given component might consider component spacing, maximum net length, and component height restrictions[59] at the same time.

P-CAD 2002 displays the board regions satisfying the desired rules, helping you determine which design constraints are limiting placement options. Components can be clustered[60] by different properties including placement side, reference designator, type, pin count, package size and height. These flexible functions assist you in organizing components for efficient placement, no matter what kind of board you are creating.

Placement highlights:
- Ability to add/remove net nodes.

[54] *n.* 定向(方位)

[55] *adj.* 参数的

[56] *n.* 焊料, 接合物

[57] *adv.* 手工地

[58] *n.* 约束, 限制

[59] *n.* 限制, 约束

[60] *vt. & vi.* 使成群, 把……集成一组

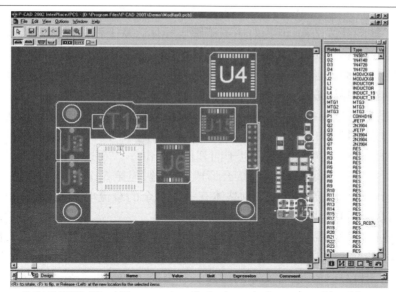

P-CAD 2002's Visual Placement Area display instantly shows you valid positions for the component you are placing, concurrently considering physical, electrical, and manufacturing design constraints.

• Island removal based on interior[61] placement, size, or sub-selection. • Move components by reference designator, with support for multiple selections. • Direct selection of highlighted objects. • Object sub-selection to access and edit reference designators, type and value attributes, and copper pour islands. • Visual[62] placement areas graphically display where a component can be placed without violating the defined.	[61] *adj.* 内部的，内的 [62] *adj.* 可见的，可视的

参考译文　基本半导体晶体结构

要了解二极管、晶体管和其他半导体设备的性能，首先要了解所有半导体设备的基本结构。早期的半导体都是由锗元素构成的，但在现代的应用中首先选用硅元素。

纯硅的晶体结构是三维的，但它很难显示或被看到，所以左图（图略）经常被用来表示硅的晶体结构。按照人们在物理学中的分类，硅（和锗）属于元素周期表中的Ⅳa列，属于碳族元素。该族元素的基本特征是每个原子都有4个电子与相邻的原子共用电子对而结合。

说简单一点儿，就是两个硅原子结合的属性是每个原子提供一个电子同另一个共享。两个电子的共享实际上就是被两个原子同等地共享。这种共享的类型就是熟知的共价键。这种键非常稳定，使得两个原子结合得非常牢固，因此需要很大的能量才能破坏这种结合。

对于那些对此感兴趣的人，还可以发现三维硅晶体以相同的键角彼此结合。如果你想象四面体的中心有一个原子（三个点在底部、第4个点在上部的一个棱锥），那么4个键将指向四面体的各个顶点。

现在我们已经了解了硅的晶体结构,但我们还不了解半导体。从上面的晶体结构我们看到,所有硅原子的最外层电子被用来同别的原子建立共价键。没有电子可以从一个地方流到另一个地方而形成电流。因此,纯硅晶体是非常好的绝缘体。事实上,几乎所有的玻璃都是由二氧化硅构成的。纯硅晶体被称为本征晶体。

为了让硅晶体导电,我们必须寻找一种方法让一些电子在晶体内移动,尽管原子之间有共价键。这里介绍一种实现方法,如右图(图略)所示:在晶体结构中掺入砷或磷等杂质。这些元素在元素周期表的Ⅴa列,最外层有5个电子同其他原子共享。在这种应用中,同前面一样,5个电子中的4个电子同相邻的硅原子结合形成共价键,但第5个电子不能结合,当加上很小的电压时这个电子很容易移动。因为晶体有一个多余的载流电子,所以每个晶体有一个负电荷,这就是众所周知的"N型"硅。

这种结构不能像铜或银那样容易导电,它对电流有一定的阻抗,称它为导体不合适,但同时它又不是绝缘体,因此,它就称为半导体。

虽然这种效果很有趣,但它自身并没有实际的用途。一个简单的碳电阻器的制造比起硅半导体要简单和便宜。我们仍然没有任何方法实际控制电子流。

但是,稍等片刻!我们通过将包含5电子的原子杂质掺入4电子原子矩阵获得了半导体(对于物理类型,我们只考虑外围电子能否形成共价键,至于内核的电子不包括在这个过程或讨论范围)。如果我们用别的方法,比如掺入3电子的杂质到这样的晶体会发生什么情况?假设,如图(图略)所示,掺入一些铝(在元素周期表中处于ⅢA列)到晶体中,情况会怎么样?我们也可以试试镓,它也处在元素周期表的ⅢA列,正好在铝的下面,情况又怎样?

这些元素只有三个电子可以同别的原子共享。这三个电子的确同相邻的硅原子形成共价键,但所希望得到的第4个共价键不能形成。由于不能形成一个完整的结合,在晶体结构中就留下一个"空穴"。

实验显示,如果有一个空位置是电子理论上应进入的,电子常常试图移入来填补空位置。然而,填补空穴的电子必须离开一个共价键去填补空位置,因此在它转移后又留下新的空穴。而后面的电子移进这个空穴,留下另一个空穴,如此继续。以这种方式,空穴看起来好像是正电荷流过晶体。因此,这种类型的半导体材料被称为"P型"硅。

就它们自身而言,P型半导体并不比N型半导体更有用。真正有趣的现象是,在一个硅体中将两种半导体以不同方式结合起来。最基本和明显的组合方式是将一个硅体中一端作成N型区,另一端作成P型区。如上描述的有两个区的硅体被称为半导体二极管。

Unit 5

Text A Number Systems

PLC is a computer. It stores information in the form of On or Off conditions (1 or 0), referred to as binary digits (bits). [1] Sometimes binary digits are used individually and sometimes they are used to represent numerical values.

1. Decimal System

Various number systems are used by PLCs. All number systems have the same three characteristics: digits, base, weight. The decimal system, which is commonly used in everyday life, has the following characteristics.

Ten digits: 0,1,2,3,4,5,6,7,8,9
Base: 10
Weights: 1,10,100,1000,…

2. Binary System

The binary system is used by programmable controllers. The binary system has the following characteristics.

Two digits: 0,1
Base: 2
Weights (Powers of base 2): 1,2,4,8,16,…

In the binary system 1s and 0s are arranged into columns. Each column is weighted. The first column has a binary weight of 2^0. This is equivalent to a decimal 1. This is referred to as the least significant bit. The binary weight is doubled with each succeeding column. The next column, for example, has a weight of 2^1, which is equivalent to a decimal 2. The decimal value is doubled in each successive column. The number in the far left hand column is referred to as the most significant bit. In this example, the most significant bit has a binary weight of 2^7. This is equivalent to a decimal 128.

128	64	32	16	8	4	2	1
0	0	0	1	1	0	0	0

3. Converting Binary to Decimal

The following steps can be used to interpret a decimal number from a binary value.

(1) Search from least to most significant bit for 1^s.

(2) Write down the decimal representation of each column containing a 1.

(3) Add the column values.

In the following example, the fourth and fifth columns from the right contain a 1. The decimal value of the fourth column from the right is 8, and the decimal value of the fifth column from the right is 16. The decimal equivalent of this binary number is 24. The sum of all the weighted columns that contain a 1 is the decimal number that the PLC has stored.

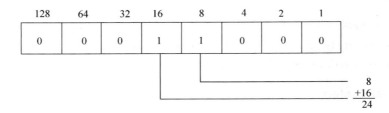

If the fourth and sixth columns from the right contain a 1. The decimal value of the fourth column from the right is 8, and the decimal value of the sixth column from the right is 32. The decimal equivalent of this binary number is 40.

4. Bits, Bytes, and Words

Each binary piece of data is a bit. Eight bits make up one byte. Two bytes, or 16 bits, make up one word.

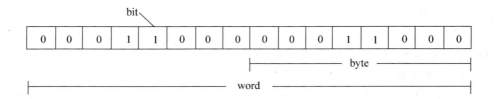

5. Logic 0, Logic 1

Programmable controllers can only understand a signal that is On or Off. The binary system is a system in which there are only two numbers, 1 and 0. Binary 1 indicates that a signal is present, or the switch is On. Binary 0 indicates that the signal is not present, or the switch is Off.

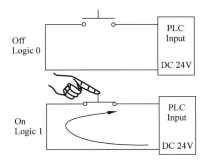

6. BCD

Binary-Coded Decimal (BCD) are decimal numbers where each digit is represented by a four-bit binary number. BCD is commonly used with input and output devices. A thumb wheel switch is one example of an input device that uses BCD. The binary numbers are broken into groups of four bits, each group representing a decimal equivalent. A four-digit thumbwheel switch, like the one shown here, would control 16(4×4) PLC inputs.

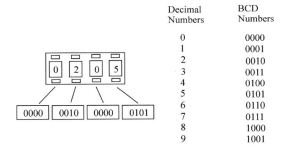

7. Hexadecimal

Hexadecimal is another system used in PLCs. The hexadecimal system has the following characteristics.

16 digits: 0,1,2,3,4,5,6,7,8,9,A,B,C,D,E,F

Base: 16

Weights (Powers of base 16): 1,16,256,4 096 …

The ten digits of the decimal system are used for the first ten digits of the hexadecimal system. The first six letters of the alphabet are used for the remaining six digits.

A = 10 D = 13
B = 11 E = 14
C = 12 F = 15

The hexadecimal system is used in PLCs because it allows the status of a large number of binary bits to be represented in a small space such as on a computer screen or programming device display. [2] Each hexadecimal digit represents the exact status of four binary bits. To convert a decimal number to a hexadecimal number the decimal number is divided by the base of 16. [3] To convert decimal 28, for example, to hexadecimal:

$$\begin{array}{r}1\\16\overline{)28}\\\underline{16}\\12\end{array}$$

Decimal 28 divided by 16 is 1 with a remainder of 12. 12 is equivalent to C in hexadecimal. The hexadecimal equivalent of decimal 28 is 1C. The decimal value of a hexadecimal number is obtained by multiplying the individual hexadecimal digits by the base 16 weight and then adding the results.[4] In the following example the hexadecimal number 2B is converted to its decimal equivalent of 43.

$$16^0 = 1$$
$$16^1 = 16$$
$$B = 11$$

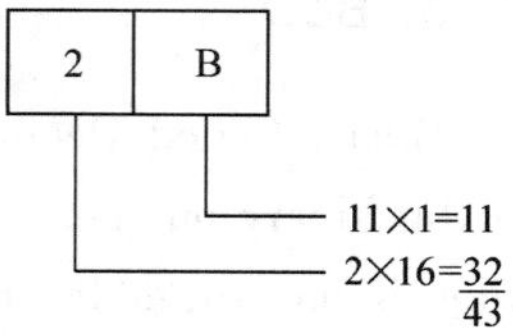

2 B

11×1=11
2×16=32
43

8. Conversion of Numbers

The following chart shows a few numeric values in decimal, binary, BCD, and hexadecimal representation.

Decimal	Binary	BCD	Hexadecimal
0	0	0000	0
1	1	0001	1
2	10	0010	2
3	11	0011	3
4	100	0100	4
5	101	0101	5
6	110	0110	6
7	111	0111	7
8	1000	1000	8
9	1001	1001	9
10	1010	00010000	A
11	1011	00010001	B
12	1100	00010010	C
13	1101	00010011	D
14	1110	00010100	E
15	1111	00010101	F
16	10000	00010110	10
17	10001	00010111	11
18	10010	00011000	12
19	10011	00011001	13
20	10100	00100000	14
⋮	⋮	⋮	⋮

New Words and Phrases

number system		数字系统
numerical value		数字值
numerical	[nju(:)'merikəl]	adj. 数字的,用数表示的
decimal	['desiməl]	adj. 十进制,十进的,小数的,以十为基础的
		n. 小数
decimal system		十进制系统
digit	['didʒit]	n. 阿拉伯数字
base	[beis]	n. 底部,基础,根据地,基地,本部,基数
		vt. 以……作基础,基于……
		adj. 低级的
weight	[weit]	n. 权重,位权,重量,分量
power	['pauə]	n. [数]幂
binary	['bainəri]	adj. 二进制的,二进位的,二元的
binary system		二进制系统
equivalent	[i'kwivələnt]	adj. 相等的,相当的,同意义的
		n. 等价物,相等物
bit	[bit]	n. 位,比特
byte	[bait]	n. (二进制的)字节,位组
word	[wə:d]	n. 字,词
succeeding	[sək'si:diŋ]	adj. 以后的,随后的
logic 0		逻辑0
logic 1		逻辑1
programmable controller		可编程控制器
BCD		二进制编码的十进制
thumbwheel switch		指轮开关
hexadecimal	[heksə'desim(ə)l]	adj. 十六进制的
		n. 十六进制
multiply	['mʌltiplai]	vt. & vi. 乘,增加
remainder	[ri'meində]	n. 余数
		adj. 剩余的
conversion of number		数字转换

Notes

[1] PLC is a computer. It stores information in the form of On or Off conditions (1 or 0), referred to as binary digits (bits).

本句涉及的是一个词组 refer to as sth(称某物为……)。完整的句子应该是:It stores

information in the form of On of Off conditions (1 or 0), which is referred to as binary digits (bits). 本句省略了连词 which，因此 is 也省略。

[2] The hexadecimal system is used in PLCs because it allows the status of a large number of binary bits to be represented in a small space such as on a computer screen or programming device display.

本句是由 because 连接的因果句。在 because 从句中，主语为 it，谓语是 allow…to…，此外还有一个词组 such as 举例说明是什么样的 small space。结构清楚后，这句话虽然看起来很长，实际上理解起来也不困难。

[3] To convert a decimal number to a hexadecimal number the decimal number is divided by the base of 16.

英语里的"加减乘除"分别是：add（加），minus（减），multiply（乘），divide（除）。例如：
If you add 4 to 3 you get 7.（4 加 3 得 7。）
12 minus 7 leaves 5.（12 减去 7 剩 5。）
3 multiplied by 5 is 15.（3 乘 5 等于 15。）
Divide eight by two you get four.（用 2 除 8 得 4。）

[4] The decimal value of a hexadecimal number is obtained by multiplying the individual hexadecimal digits by the base 16 weight and then adding the results.

本句是一个简单句。主语是 the decimal value，谓语是 is obtained by，怎样得到是通过 multiply（相乘）以及 add（相加）。也就是说，by 后包括两个动词，multiply 以及 add，而不仅是 multiply。所以本句话理解为，十六进制数对应的十进制的值是将十六进制的各位分别乘对应的位权，然后将它们的结果相加。

Exercises

【Ex. 1】 根据课文内容，回答以下问题。

（1）What form is used to store information in PLCs?

（2）What characteristics do all number systems have?

（3）Which number system is commonly used in everyday life?

（4）What characteristics do binary system have?

（5）What is the decimal number for the hexadecimal number 3C?

【Ex. 2】 根据下面的英文解释,写出相应的英文词汇。

英 文 解 释	词　　汇
knowledge of a specific event or situation; intelligence	
of or relating to a number or series of numbers	
the combination of qualities or features that distinguishes one person, group, or thing from another	
of or relating to a system of numeration having 2 as its base	
a number written using the base 10; numerical	
of, relating to, or based on the number 16	
binary coded decimal	
the letters of a language, arranged in the order fixed by custom	
the number that is raised to various powers to generate the principal counting units of a number system	
the number left over when one integer is divided by another	

【Ex. 3】 把下列句子翻译为中文。

(1) Because a single bit can only store two values, bits are combined together into large units in order to hold a greater range of values.

(2) Boolean variables use a single bit to hold their value, so can only assume one of two possible states.

(3) Consider where a computer allocates 16 bits of storage per status variable. If we had three status variables, the space consumed would be 48 bits.

(4) ASCII is a computer code which uses 128 different encoding combinations of a group of seven bits ($2^7 = 128$) to represent.

(5) Characters are non-numeric symbols used to convey language and meaning. In English, they are combined with other characters to form words.

(6) Computer system normally stores characters using the ASCII code. Each character is stored using eight bits of information, giving a total number of 256 different characters ($2^8 = 256$).

(7) Text strings are a sequence of characters (words or multi- character symbols). Each character is stored one after another, each occupying 8 bits of memory storage.

(8) Numeric information cannot efficiently be stored using the ASCII format. Imagine storing the number 123 769 using ASCII. This would consume 6 bytes, and it would be difficult to tell if the number was positive or negative (though we could precede it with the character + or −).

(9) The computer industry agreed upon a standard for the storage of floating point numbers. It is called the IEEE 754 standard, and uses 32 bits of memory (for single precision), or 64 bits (for double precision).

(10) An array is a group of elements which are all of the same type, size and name. It can be thought of as a box with multiple compartments, each compartment capable of storing one data item.

【Ex. 4】 把下列短文翻译成中文。

There are two problems with integers; they cannot express fractions, and the range of the number is limited by the number of bits are used. An efficient way of storing fractions is called the floating point method, which involves splitting the fraction into two parts, an exponent and a mantissa. The exponent represents a value raised to the power of 2. The mantissa represents a fractional value between 0 and 1.

【Ex. 5】 通过Internet查找资料，借助如"金山词霸"等电子词典和辅助翻译软件，完成以下技术报告。通过E-mail发送给老师，并附上你收集资料的网址。

在Internet上查找关于进位记数制和数字系统的基本知识。

Text B Digital Circuit Elements

1. CMOS Element and Watch Switching

The complementary MOSFET scheme (or CMOS) started the second revolution in computational machines. The limits of speed and density were conquered by the move to semiconductors and Very Large Scale Integration, but the power consumption and circuit cooling demands of bipolar transistors packed at extreme densities were formidable problems. The problem is that the transistors were always "ON" (in other words drawing current and dissipating energy). CMOS circumvents this problem and allows bits to be stored without constant power consumption. A schematic of the CMOS inverter is given in the figure below. As discussed in class the device dissipates energy only when it is switched from high to low or low to high. Quiescent operation in either the high or the low state dissipates essentially no power. So cooling the circuit is much easier, and supplying power is much less problem. If you don't believe me, just ask your calculator, digital watch or your laptop.

Connect $V_{DD} = +5V$ and ground to the CD4007 pins as depicted below using only one set of transistors. For example, pin 10 = V_I, pin 11 = V_{DD}, pin 9 = GND and pin 12 = V_O. Connect a 500Ω resistor between V_{DD} and pin 11 for better performance.

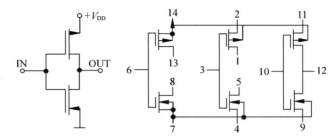

Slowly ramp the input voltage from zero up to 3.5V. At some point the output should switch from high to low. Note the voltage when the switch occurs.

Now connect a 100Ω resistor in series with pin 9 above the ground point.

Try to measure the transient current (momentary voltage across the resistor) as you slowly ramp the input voltage up and down to make the output switch. If you can't see the signal you can cheat by using the Miller effect by adding a medium sized capacitor between output and ground.

Try to measure the intrinsic switching time and estimate the power consumption for such an inverter switched at 1MHz compared with a bipolar circuit where the devices are constantly passing current.

Estimate the power consumption for switching at 1GHz.

2. Gates, Truth Tables, and Pull-up Resistors

One of the simplest gates is the inverter. The Boolean equation for the inverter is
$$Y = \bar{A}$$
The following is the diagrammatic representation of the inverter.

The 7404 chip contains 6 inverters and can be schematically represented as follows

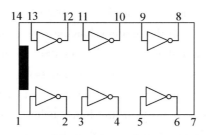

The diagram of the chip is drawn as if you are viewing it from above. Note the thick black line that is used as an orientation mark, located at the left end of the chip. Also note that pins 14 and 7 are not connected to an inverter, they are the power supply connections for the chip. Pin 14 must be held at +5V with respect to pin 7. No pin may held at a voltage greater than that at pin 14 or less than that at pin 7. So if pin 14 is held at +5V then nothing can be greater than +5V and if pin 7 is grounded then you cannot have a voltage to the chip that is less than 0V. This supply pin assignment is common for 7400 series TTL. If you have a question about the wiring of a particular chip, refer to the TTL cookbook; a copy will be kept in the lab. You may also use the websites linked to the course homepage.

3. Procedure

(1) Wire a 7404 inverter, on the Digital-Designer as follows

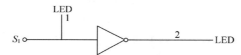

(2) Apply a clock signal from the clock on the Digit-Designer Box to the input of one of the inverters in the 7404 IC chip. Simultaneously look at the input and output on the oscilloscope and also the LED's on the Digital-Designer for a range of clock frequencies from 1kHz to 100kHz.

Comment on the input and output of the circuit. Are there any timing problems with this circuit?

(3) Wire 6 inverters from the 7404 IC chip in series. Connect the Digital-Designer clock to the first inverter.

Observe the input of the 1st and the output of the last inverter simultaneously on the oscilloscope. Determine the "Propagation delay" through a single inverter.

(4) Wire the following circuit using an open-collector inverter (7404). The 2kΩ. "pull-up" resistor is necessary for speed and noise immunity when driving a TTL input. The numbers that you see on the diagram are to distinguish the input pins on the 7404 IC chip. So the 1 means that the input A should be connected to pin 1 on the 7404 IC chip and so on. The equation at the bottom right is the algebraic representation of the logical NOR equation.

Verify that the circuit performs the logical NOR function.

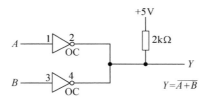

How does the inverter output differ from an inverting Op-amp that you worked with in Lab 6?

4. Translating Boolean Equations Into Electronic Circuits

In many cases, translation of Boolean equations into electronic logic can be accomplished by a straightforward, one-for-one replacement of a term or group of terms in the equation. As an illustration, we will consider the 4-input data selector or "multiplexer".

We have 4 digital signals that we would like to be able to send over a single wire. We need logic that defines the output Y of a circuit to be the nth input signal X_n, where the number n is given. Since n can take on 4 values, it must be a 2-digit binary number, which we will call BA. In other words, BA selects the input X_n, which will appear on the output. The truth table expressing this circuit is:

B	A	Y(output)
0	0	X_0
0	1	X_1
1	0	X_2
1	1	X_3

From this table, a Boolean equation can be written by expression:
$$Y = (\overline{B} \cap \overline{A} \cap X_0) \cup (\overline{B} \cap A \cap X_1) \cup (B \cap \overline{A} \cap X_2) \cup (B \cap A \cap X_3)$$

From the Boolean equation above, the following schematic that describes the equation can be drawn.

In implementing the logic displayed in this diagram, we are slightly hampered by the fact that a 7400 series 4-input OR gate does not exist. The elegant solution to this problem involves DeMorgan's theorem and some common sense.

Show that the circuit above is equivalent to the one below by writing the Boolean equations for both and using DeMorgan's Theorem to transform one into another.

Build the above circuit on your Digital-Designer and write out the truth table for the circuit. (Isn't this a real pain to build? Now imagine building the Pentium with its 50 million switches. That's why computers do all of the layout and wiring in foundries.)

Verify that your circuit is working properly for four combinations in your truth table.

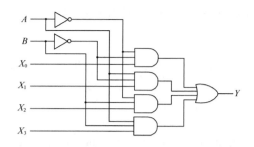

New Words and Phrases

CMOS (Complementary Metal-Oxide-Semiconductor)		n.	互补金属氧化物半导体
complementary	[ˌkɔmplə'mentəri]	adj.	补充的，补足的
MOSFET (Metal Oxide Semiconductor Field-effect Transistor)	['mɔsfet]		金属氧化物半导体场效应晶体管
scheme	[skiːm]	n.	安排，配置，计划，方案，图解，摘要
		vt. & vi.	计划，设计，图谋，策划
computational	[ˌkɔmpju(ː)'teiʃ(ə)n(ə)l]	adj.	计算的
density	['densiti]	n.	密度
integration	[ˌinti'greiʃən]	n.	集成
Very Large Scale Integrated circuit			超大规模集成电路
bipolar	[bai'pəulə]	n.	双极型晶体管
formidable	['fɔːmidəbl]	adj.	强大的，令人敬畏的，艰难的
circumvent	[ˌsəːkəm'vent]	vt.	围绕，包围，智取
inverter	[in'vəːtə]	n.	反相器
quiescent	[kwai'esənt]	adj.	静止的
laptop	['læptɔp]	n.	便携式电脑，膝上型电脑
pin	[pin]	n.	钉，销，栓，大头针，别针，腿
		vt.	钉住，别住，阻止，扣牢，止住，牵制
depict	[di'pikt]	vt.	描述，描写
ramp	[ræmp]	n.	斜坡，坡道
		vi.	蔓延
		vt.	使有斜面
momentary	['məumənteri]	adj.	瞬间的，刹那间的
inverter switch			换向开关
Boolean	['buːliən]	n.	布尔

		adj.	布尔的
gate	[geit]	*n.*	逻辑门
truth table			真值表
pull-up resistor			上拉电阻
diagrammatic	[,daɪəgrə'mætik]	*adj.*	图表的,概略的
orientation	[,ɔ(:)rien'teiʃən]	*n.*	方向,方位,定位,倾向性,向东方
chip	[tʃip]	*n.*	碎片,芯片
		vt. & *vi.*	削成碎片,碎裂
website			WWW(万维网)的站点
immunity	[i'mju:niti]	*n.*	免疫性
translation	[træns'leiʃən]	*n.*	转化,转换,平移,翻译,译文
multiplexer	['mʌlti,pleksə]	*n.*	多路(复用)器
theorem	['θiərəm]	*n.*	定理,法则

Exercises

【Ex. 6】 根据文章所提供的信息判断正误。

(1) The CMOS started the second revolution in computational machines.

(2) Now, the limits of speed and density are formidable problems.

(3) CMOS can store bit without constant power consumption.

(4) Quiescent operation in either high or low state dissipates a little energy.

(5) The 7404 chip contains 6 inverters.

(6) None of the pins of the 7404 chip is connected to an inverter.

(7) Pin 14 and pin 7 from the 7404 chip are the power supply connections for the chip.

(8) For the 7404 chip, every pin may be held at a voltage greater than pin 7.

(9) If pin 14 is held at a voltage +24V and pin 7 is held at a voltage +0V, the 7404 chip will work well.

(10) For the 7404 IC chip, pin 14 must be held at +5V with respect to pin 7. No pin may be held at a voltage greater than that at pin 14 or less than that at pin 7.

科技英语翻译知识　词类的转换

英汉两种语言在词汇的使用上有很大的不同。汉语里多使用动词,英语,尤其是科技英语里多使用名词;英语有许多名词派生的动词,以及由名词转用的动词,而汉语里没有;汉语里形容词可作谓语,而英语里形容词只能用做定语等。因此在英译汉过程中,就不可以逐词死译,而需要转换词性,才能使汉语译文通顺自然。

词性转换分为以下4种情况。

1. 转换成汉语动词

(1) Reversing the direction of the current reverses the direction of its lines of force.

倒转电流的方向也就倒转了它的磁力线的方向。

(动名词 reversing 转换成动词"倒转"。)

(2) It is desirable to transform high voltage into low voltage by means of a transformer in an AC system.

我们希望通过交流系统中的变压器把高压变为低压。

(形容词 desirable 转换成动词"希望"。)

(3) The electric current flows through the circuit with the switch on.

如果开关接通,电流就流过线路。

(副词 on 转换成动词"接通"。)

(4) The letter E is commonly used for electromotive force.

通常用字母 E 表示电动势。

(介词 for 转换成动词"表示"。)

2. 转换成汉语名词

(1) Neutrons act differently from protons.

中子的作用不同于质子。

(动词 act 转换成名词"作用"。)

(2) The electrons flow from the negative to the positive.

电子从负极流向正极。

(形容词 the negative 转换成了名词"负极"。)

(3) Such magnetism, because it is electrically produced, is called electromagnetism.

由于这种磁性产生于电,所以称为电磁。

(副词 electrically 转换成名词"电"。)

(4) The voltage induced in the core is small, because it is essentially a winding having but one turn.

铁芯的感应电压小,因为铁芯实质上是一只有一匝的绕组。

(代词 it 转换成名词"铁芯"。)

3. 转换成汉语形容词

(1) Thus, the addition of the inductor prevents the current from building up or going down quickly.

这样,添加的电感器阻止电流不能迅速增加或下降。

(为与谓语"阻止"搭配,将定语"电感器"转译为主语,而将名词主语"添加"转换为"电感器"的定语,变为形容词。)

(2) Gases conduct best at low pressure.

气体在低压下导电性能最佳。

(副词 best 转换为形容词"最佳"。)

(3) The electric conductivity has great importance in selecting electrical materials.

导电性在选择电气材料时很重要。

(名词 importance 转换成形容词"重要"。)

(4) Why is electricity widely used in industry?

为什么电在工业中得到广泛的使用呢?

(动词 use 转换成名词,所以副词 widely 相应地转换成形容词"广泛的"。)

4．转换成汉语副词

(1) The potentials of both the grid and plate are effective controlling plate current.

栅极和屏极两者的电位有效地控制着屏极电流。

(形容词 effective 转译成副词"有效地"。)

(2) The added device will ensure accessibility for part loading and unloading.

增添这种装置将保证工件装卸方便。

(名词 accessibility 转换成副词"方便"修饰动词"装卸"。)

(3) In the long and laborious search, Marie Curie succeeded in isolating two new substances.

在长期辛勤的研究中,玛丽·居里成功地分解出了两种新物质。

(动词 succeed 转译成副词"成功地"。)

(4) In human society activity in production develops step by step from a lower to higher level.

人类社会的生产活动,由低级向高级逐步发展。

(状语 step by step 转译成副词"逐步"修饰发展。)

Reading Material

阅读下列文章。

Text	Note
P-CAD 2002 Overview(2) **1. P-CAD 2002 Design Synchronizer** 　　P-CAD 2002's schematic editor provides advanced linking between your schematic and PCB layout, allowing you to transfer design data, automatically cross-probe between documents, and forward and back—annotate your design. 　　P-CAD's netlist supports system and user-defined attributes[1] as well as component and net connectivity. The netlist includes all design constraints[2] and attributes, including those that are defined mathematically based on other constraints. These are subsequently used to constrain PCB placement, routing and manufacturing. 　　P-CAD 2002 uses advanced DDE (Dynamic Data Exchange) techniques to enable a variety of cross-probe features. When both a schematic and its	[1] *n*. 属性,品质,特征 [2] *n*. 约束,强制,局促

corresponding[3] PCB documents are open on the desktop, highlighted component and net information are visible in both applications. For example, highlighting a net or component in one document highlights the corresponding net or component in the other.

A full set of Engineering Change Orders (ECOs) is supported for forward and back annotation with the PCB layout to keep them synchronized and current. From the schematic you can record ECOs for use by the PCB editor, or import ECOs generated from the PCB, depending on where the change was initiated. Pending ECOs can be viewed while still in memory, giving you full control of the forward and back annotation process.

ECOs can record reference designator changes (Was-Is), net name changes, pin and gate[4] swap changes, and additions, deletions, and modifications of components, parts, nets, net nodes, and attributes.

Design synchronization feature highlights:
- "One button" design synchronization.
- Transfer design data between schematic and PCB environments, automatically cross-probe[5] between documents, and forward and back-annotate your design.
- Ability to view context-sensitive properties of all items.
- Extensive electrical rules checking with many techniques to manage errors.

2. P-CAD 2002 Manual Routing

For precise control over track placement P-CAD 2002 includes a host of powerful manual routing features that let you quickly lay down traces[6] exactly where you want.

Manually route traces using 45°, 90° and any angle curved[7] placement modes, with hot-key control over corners. Corners can be constrained to 45° or orthogonal[8], or placed at any angle. Corner arcs are fully supported as well. Any-angle, tangential[9] arc routing supports high speed and flex board requirements, with the arc radius specified on the status line.

Easily start a route anywhere along existing copper using P-CAD 2002's "T" and "Y" routing features. Change layers during manual routing and P-CAD 2002 will automatically insert a via for you. Trace width and via style attributes are automatically applied.

While routing a connection you can stop placing traces at any point and let P-CAD 2002 complete the connection for you, or leave the connection partially routed with the option to automatically re-optimize connection lines to the end of the partial route. Manually route to the center of a pad[10], via, or the endpoint of a line that is not part of any net and P-CAD 2002 adds that item to the net being routed.

Manual Routing highlights:
- Any angle, 45°, 90°, and curved placement modes.
- Manual routing uses net attributes for trace width and via style.
- Manual routing info on Status line: trace length, net name, ortho mode.
- Auto via insertion when changing layers during manual routing.
- Backtrack and redundant trace cleanup[11], and loop removal.

[3] *adj.* 相应的

[4] *n.* (逻辑)门

[5] *n.* 交叉索引

[6] *n.* 痕迹,迹线,路径
[7] *adj.* 弯曲的
[8] *adj.* 直角的,直交的
[9] *adj.* 切线的

[10] *n.* 垫,衬垫

[11] *n.* 清除

- Y- and T-routing.
- 90° arc, 45° corner, and Y- and T-junction mitering.
- Net being routed is highlighted[12].
- Optimize unrouted[13] connections.
- Can add additional vertices to existing lines and traces.

3. P-CAD 2002 Autorouting[14]

Designed to work hand-in-hand with the PCB editor, P-CAD 2002's shape-based autorouter is based on the latest in routing algorithms[15] and produces high-quality, fast autorouting with high completion rates.

[12] *adj.* 突出的
[13] *adj.* 不成回路的
[14] *n.* 自动布线
[15] *n.* [数]运算法则

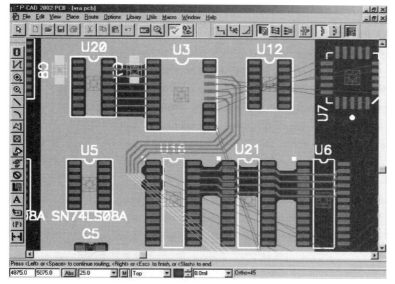

Powerful interactive routing features such as bus routing, multi-wire routing, push routing and the ability to calculate and display the available routing area make light work of even the toughest routing jobs.

The autorouter includes heuristic[16] routers, a two-layer power and ground router, a memory router, a dispersion[17] or fan-out router, a family of different pattern routers, a push-and-shove routing algorithm, a rip up algorithm, a manufacturing pass which spreads tracks to equalize[18] spacing once routing is complete, and a variety of cleanup routines that give high-quality finished routing with less re-work. And because the router is shape-based, it easily handles both through-hole and the latest surface mount designs.

The autorouter uses complex polygonal shapes to determine available routing areas. The use of polygonal shapes makes maximum use of board area and allows true diagonal routing with definable routing directions and angles for each board layer. As well, the autorouter uses adaptive costings to maximize completion rates and route quality for each board. Before routing, the design is automatically analyzed and an appropriate cost model selected. During routing, the cost model is intelligently updated when necessary to ensure high completion rates regardless of board complexity or density.

[16] *adj.* 启发式的
[17] *n.* 散布,散射
[18] *vt.* 使相等,补偿

Autorouting highlights:
- Shape-based autorouter based on the latest in routing algorithms to produce high-quality, fast autorouting with high completion rates.
- Easily handles both through-hole and the latest surface mount designs.
- Use of polygonal shapes allows true diagonal[19] routing with definable routing directions and angles for each board layer.
- Adaptive costings used to maximize completion rates and route quality for each board.
- Full interface to Cadence's SPECCTRA® interactive and batch routing products.

4. P-CAD 2002 Design-For-Manufacture support

P-CAD 2002 gives you full design output and documentation support to ensure a smooth passage to manufacture. With extensive DXF import/export support, full Gerber output control, and complete design annotation and documentation features P-CAD 2002 meets all your output needs. From within the PCB editor, comprehensive dimension[20] editing features allow you to effortlessly move dimensions and edit properties. Extensive DXF import/export support, full Gerber output control, and comprehensive design annotation and documentation features ensure a smooth passage to manufacture.

P-CAD 2002's IDF translator ensures a fluid[21] exchange of design data between the P-CAD PCB editor and mechanical designers, ensuring clear communications for more effective solid model form-fit analysis and placement of critical components.

For more efficient design for manufacture, P-CAD 2002 includes direct support for revolutionary WebQuote system, allowing you to request a quotation to fabricate or assemble your board from within the PCB editor. Using the internet, WebQuote, links you with PCB fabricators[22] worldwide and allows you to easily seek quotes tailored to your specific criteria.

Whether you're documenting your design, generating manufacturing files, or publishing your library data, P-CAD 2002 is there for all your manufacturing output and documentation needs.

Design-for-manufacture feature highlights:
- "One click" output generation.
- Supported output types include Gerber files, NC Drill files, Bill Of Materials, DRC/ERC errors, library contents and custom reports.
- Custom reports allow field selection and selection criteria[23] plus sorting options.
- Component library documentation.

5. P-CAD 2002 Signal integrity analysis

Ensure the "real world" performance of your board design with P-CAD 2002's sophisticated signal integrity simulator. Providing detailed and valuable information on

[19] adj. 斜的,斜纹的,对角线的

[20] n. 尺寸,尺度,维(数),度(数),元

[21] adj. 不固定的,可改变的

[22] n. 制作者

[23] n. 标准,准则

overshoot, undershoot, impedance[24] and signal slope, the simulator takes the guesswork[25] out of your design.

To investigate problems, P-CAD 2002's integrated signal integrity simulator uses advanced transmission line calculations and I/O buffer macro-model information as input for accurate signal integrity analysis of your board using a fast reflection and crosstalk simulator model. A powerful CRO-like wave analyzer shows simulation results of any net in graphical form, with full zoom and measurement support.

P-CAD 2002's signal integrity simulator includes a powerful Termination Advisor that lets you explore the effects of different termination options without having to alter your board. Simply select a termination method for any node in the Advisor and re-run the analysis to see the results. Don't waste time with trial-and-error board redesigns, with P-CAD 2002 you can eliminate signal integrity problems before committing[26] your board to manufacture.

Signal integrity feature highlights:
- Automatically takes net information from the PCB design.
- Fast simulation of reflection and crosstalk effects on selected nets.
- No knowledge of SPICE or analog simulation is required as the Signal Integrity Simulator uses I/O buffer macro-models.
- Powerful wave analyzer to display results of reflection and crosstalk simulations.
- Powerful Termination Adviser for exploring various termination options.

[24] n. [电]阻抗,全电阻
[25] n. 凭猜测所做的工作
[26] vt. 把……交托给,提交

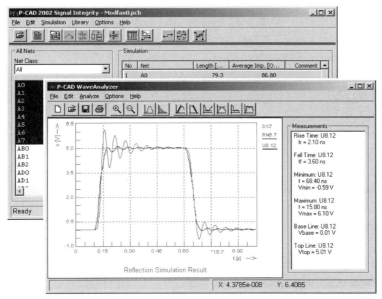

The Signal Integrity Simulator includes a powerful wave analyzer that graphically displays the results of reflection and crosstalk simulations and allows you to take measurements directly from the displayed waveforms.

Signal integrity feature highlights:
- Automatically takes net information from the PCB design.
- Fast simulation of reflection and crosstalk[27] effects on selected nets.
- No knowledge of SPICE or analog simulation is required as the Signal Integrity Simulator uses I/O buffer[28] macro-models.
- Powerful wave analyzer to display results of reflection and crosstalk simulations.
- Powerful Termination Adviser for exploring various termination options.

6. P-CAD 2002 Library management

Finding, creating and managing components are fundamental to efficient design, and P-CAD 2002 is specifically designed to put every aspect of library management under your complete control.

P-CAD 2002 takes library organization to new heights with a powerful library manager that acts as your complete library control center. P-CAD 2002's unique integrated component library system stores all components information, making true forward/back annotation between the schematic and PCB possible. Create integrated components complete with schematic symbols and PCB patterns, copy and paste components between libraries, and link components to your corporate[29] databases.

P-CAD 2002 allows you to create and store multiple and varying pattern graphics for each component, giving you the option of manual or automatic graphic swapping to suit component orientation or specific design requirements, enhancing placement efficiency.

To save your time and resources, a comprehensive range of libraries with over 27 000 components are supplied with P-CAD 2002, with updates and additional components regularly made available from www.pcad.com. To ensure these libraries meets standards you can depend on they have been developed under an ISO 9001 accredited[30] quality assurance system, so you can be confident of the accuracy of each and every component.

Library management highlights:
- Powerful library management providing complete control and organization.
- Complete component information storage providing forward/back annotation between schematic and PCB.
- Creation of integrated components complete with schematic symbols and PCB patterns.
- Copy and paste components between libraries and link them to your corporate databases.
- Over 27 000 supplied components with updates and additional components available from this website.
- Libraries developed under ISO 9001 accredited quality assurance[31] system.

7. P-CAD 2002 Bonus Technologies!

To add even more power to your PCB Design Package, we've included two bonus[32] technologies that will further enhance your design capabilities!

[27] *n.* 干扰

[28] *n.* 缓冲器

[29] *adj.* 共同的,全体的

[30] *adj.* 可信任的,质量合格的

[31] *n.* 保证,担保

[32] *n.* 附带的优点 [好处]

8. CAMtastic DXP

Also included with P-CAD 2002 is CAMtastic DXP, giving you the power to directly export your PCB files into the perfect environment to view, print, convert, edit, analyze and verify critical manufacturing data.

CAMtastic DXP loads most popular CAM formats and provides all the graphical and geometrical tools you need to Edit/Modify any design. With extensive fabrication and assembly tools, CAMtastic DXP allows you to gracefully translate between all commonly used CAM formats. Loading, editing and outputting manufacturing data have never been easier.

9. Mixed-Signal Circuit Simulator

To achieve timely product development it's crucial[33] to know that your circuit will work just the way you expect. P-CAD 2002 now includes a powerful mixed analog and digital circuit simulator, based on the latest XSPICE 3f5 engine, allowing you to verify and fine tune your design before laying out the board.

The Mixed-Mode Circuit Simulator, supplied with each P-CAD 2002 license, uses advanced analog and digital modeling techniques to generate accurate "real-world" simulations of analog, digital and mixed-mode circuits directly from a P-CAD 2002 schematic at the touch of a button.

With no limit on the number of circuit nodes the simulator can handle designs of any complexity. Effects such as propagation delays, setup and hold times, output loading—in fact almost all physical circuit parameters—are fully accounted for, ensuring you get results you can trust. And P-CAD 2002 comes supplied with a comprehensive set of simulation-ready libraries containing almost 6 000 analog and digital devices, allowing you to model your design "straight from the box".

Being a true mixed-signal[34] environment, the Mixed-Mode Circuit Simulator integrates the simulation of continuous analog waveforms and discrete[35] digital signals. You can run and view complex analog and digital simulation waveforms side-by-side, giving a full picture of your circuit's performance.

And because the Mixed-Mode Circuit Simulator provides true SPICE compatibility—the industry standard for analog simulation—you can use all the latest device manufacturers' models directly with P-CAD 2002.

Mixed-Signal Simulation highlights:
- Integration with the schematic editor to automatically generate the schematic netlist at the press of a button.
- Uses enhanced version of SPICE 3f5/XSPICE allowing accurate simulation of any combination of analog and digital devices.
- Over 6 700 simulation-ready schematic components in more than 20 libraries.
- Digital SimCode™ (XSPICE compatible) for modeling of digital devices.
- Sophisticated component sweep[36] and Monte Carlo analysis modes.
- Simultaneous display of two different waveform types.

[33] *adj.* 至关紧要的

[34] 混合信号
[35] *adj.* 不连续的，离散的

[36] *vt. & vi.* 扫过，掠过

参考译文　数字系统

PLC 就是一种计算机。它是以 On 或 Off 的形式(1 或 0)来存储信息的,也就是二进制(位)的方式。有时二进制数单独使用,有时它们用来表示数字值。

1. 十进制系统

PLC 中使用各种数制系统。所有的数制都有相同的三个属性:数字、基数和权。日常生活中经常用到的十进制系统,具有下面的属性。

10 个数字:0,1,2,3,4,5,6,7,8,9。

基数:10。

权:1,10,100,1000,…

2. 二进制系统

可编程序控制器中使用二进制。二进制系统具有下面的属性。

两个数字:0,1。

基数:2。

基数为 2 的位权:1,2,4,8,16,…

在二进制系统中,1^s 和 2^s 排成列,每列对应一个权。第一列具有二进制权 2^0。这个值等于十进制的 1。这是最低有效位。二进制中每列紧接的下一列的权是它的两倍。例如,下一列的权是 2^1,它相当于十进制中的 2。相邻着的下一列的十进制值是这列的两倍。一个数中最左边的列意味着具有最高的有效"位"。在这个例子中,最高的有效权具有二进制位权 2^7。它的值等于十进制数 128。

(图表略)

3. 二进制到十进制的转换

下列步骤可用来说明如何将二进制数转换为十进制数。

(1)从最低到最高找到有效位 1^s。

(2)将所有各列为 1 的位权所代表的十进制都记下。

(3)将各列的值相加。

在下面的例子中,从右数第 4、第 5 个数为 1。从右数第 4 位的 1 相当于十进制数 8,从右数第 5 位的 1 相当于十进制数 16。这个二进制数相应的十进制数是 24。所有的各列为 1 的位权的和是 PLC 中存储的十进制数。

(图表略)

如果从右数第 4 和第 6 列为 1。右面第 4 位所对应的十进制值是 8,右面第 6 位所对应的十进制值是 32。这个二进制数相应的十进制数是 40。

4. 位、字节和字

二进制最小单位是一位。每 8 位组成一个字节。两个字节或 16 位组成一个字。

(图表略)

5. 逻辑 0,逻辑 1

可编程序控制器只能理解 On 或 Off 信号。在二进制系统中只有 1 和 0 两个数字。二进制 1 表示有信号或开关为 On。二进制 0 表示没有信号或开关为 Off。

(图表略)

6. BCD 码

二进制编码的十进制(BCD)是每位由 4 位二进制数表示的十进制数。输入和输出设备上通常使用 BCD 编码。拨盘开关就是利用 BCD 编码进行输入的一个例子。二进制数被分成 4 位一组,每组表示一个等值的十进制数。如图所示的 4 位拨盘开关,将控制 16(4×4)个 PLC 的输入。

(图表略)

7. 十六进制

十六进制是 PLC 使用的另一种系统,十六进制有下列属性。

16 个数字:0,1,2,3,4,5,6,7,8,9,A,B,C,D,E,F。

基数:16。

基数为 16 的位权:1,16,256,4 096,…

十进制中的 10 个数字被用做十六进制系统中的前 10 个数字,字母表中的开始 6 个字符用做十六进制系统中的其余 6 个数字。

$$A = 10 \quad D = 13$$
$$B = 11 \quad E = 14$$
$$C = 12 \quad F = 15$$

十六进制系统在 PLC 中被采用,是由于它可以把很多的二进制位的状态表示在很小的空间中,比如表示在计算机屏幕或编程器的显示屏上。每个十六进制数字可以表示 4 位二进制数字的准确值。为了将十进制数转换为十六进制数,这个数要被基数 16 除。例如,转换十进制 28 到十六进制:

(计算过程略)

十进制数 28 被 16 除,商为 1,余数为 12。12 相当于十六进制中的 C。十进制 28 对应的十六进制数是 1C。要得到十六进制数对应的十进制的值,需要将十六进制的各位分别乘对应的位权,然后将其结果相加。下面的例子是把十六进制数 2B 转换为十进制的等价值 43。

$$16^0 = 1$$
$$16^1 = 16$$
$$B = 11$$

(计算过程略)

8. 数的转换

下面的表格显示了几个数分别以十进制、二进制、BCD 和十六进制表示的值。

(图表略)。

Unit 6

Text A AC Motors

AC motors are used worldwide in many residential, commercial, industrial, and utility applications. Motors transform electrical energy into mechanical energy. An AC motor may be part of a pump or fan, or connected to some other form of mechanical equipment such as a winder, conveyor, or mixer. AC motors are found on a variety of applications from those that require a single motor to applications requiring several motors. [1]

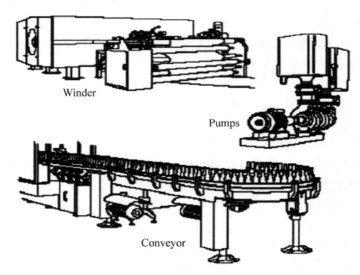

Winder

Pumps

Conveyor

1. Force and Motion

Before discussing AC motors, it is necessary to understand some of the basic terminology associated with motor operation. Many of these terms are familiar to us in some other context. Later in the course we will see how these terms apply to AC motors.

2. Force

In simple terms, a force is a push or a pull. Force may be caused by electromagnetism、gravity, or a combination of physical means.

3. Net force

Net force is the vector sum of all forces that act on an object, including friction and gravity. When forces are applied in the same direction they are added. For example, if two 10N forces were applied in the same direction the net force would be 20N.

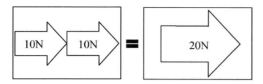

If 10N force were applied in one direction and 20N force applied in the opposite direction, the net force would be 10N and the object would move in the direction of the greater force.

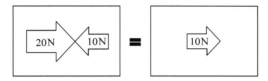

If 10N of force were applied equally in both directions, the net force would be zero and the object would not move.

4. Torque

Torque is a twisting or turning force that causes an object to rotate. For example, a force applied to the end of a lever causes a turning effect or torque at the pivot point. Torque (M) is the product of force and radius (lever distance).

$$M = \text{Force} \times \text{Radius}$$

In the English system torque is measured in pound-feet (lb-ft) or pound-inches (N-in). If 10N of force were applied to a lever 1ft long, for example, there would be 10N-ft of torque.

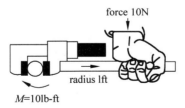

An increase in force or radius would result in a corresponding increase in torque. Increasing the radius to two ft, for example, results in 20lb-ft of torque.

5. Speed

An object in motion travels a distance in a given time. Speed is the ratio of the distance traveled and the time it takes to travel the distance.

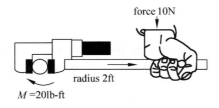

$$\text{Speed} = \frac{\text{Distance}}{\text{Time}}$$

6. Linear Speed

The linear speed of an object is a measure of how fast the object move from point A to point B. Linear speed is usually given in a form such as meters per second (m/s). For example, if the distance between point A and point B were 10 meters, and it took 2 seconds to travel the distance, the speed would be 5m/s.

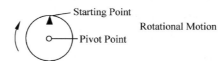

7. Angular (Rotational) Speed

The angular speed of a rotating object is a measurement of how long it takes a given point on the object to make one complete revolution from its starting point.[2] Angular speed is generally given in revolutions per minute (rad/min). An object that makes ten complete revolutions in one minute, for example, has a speed of 62.8rad/min.

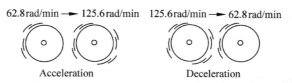

8. Acceleration

An object can change speed. The increase in speed is called acceleration. Acceleration occurs only when there is a force acting upon the object. An object can also change from a higher to a lower speed. This is known as deceleration (negative acceleration). A rotating object, for example, can accelerate from 62.8rad/min to 125.6rad/min, or decelerate from 125.6rad/min to 62.8rad/min.

9. Inertia

Mechanical systems are subject to the law of inertia. The law of inertia states that an object will tend to remain in its current state of rest or motion unless acted upon by an external force. This property of resistance to acceleration/deceleration is referred to as the moment of inertia. The English system of measurement is pound-feet squared ($lb\text{-}ft^2$). If we look at a continuous roll of paper, for example, we know that when the roll is stopped it would take a certain amount of force to overcome the inertia of the roll to get it rolling.[3] The force required to overcome this inertia can come from a source of energy such as a motor. Once rolling, the paper will continue unwinding until another force acts on it to bring it to a stop.

10. Friction

A large amount of force is applied to overcome the inertia of the system at rest to start it moving. Because friction removes energy from a mechanical system, a continual force must be applied to keep an object in motion. The law of inertia is still valid, however, since the force applied is needed only to compensate for the energy lost. Once the system is in motion only the energy required to compensate for various losses need be applied to keep the conveyor in motion.[4] In the previous illustration, for example, these losses include:
- Friction within motor and driven equipment bearings[5]
- Wind losses in the motor and driven equipment
- Friction between materials on winder and rollers[5]

11. Work

Whenever a force of any kind causes motion, work is accomplished. For example, work is accomplished when an object on a conveyor is moved from one point to another.

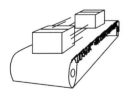

Work is generally expressed in foot-pounds (ft-lb) and is defined by the product of the net force (F) applied and the distance (d) moved. If twice the force is applied, twice the work is done. If an object moves twice the distance, twice the work is done.

$$W = F \times d$$

12. Power

Power is the rate of doing work, or work divided by time.

$$\text{Power} = \frac{\text{Force} \times \text{Distance}}{\text{Time}}$$

In other words, power is the amount of work it takes to move the package from one point to another point, divided by the time.

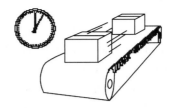

13. Horsepower

Power can be expressed in foot-pounds per second, but is often expressed in horsepower (hp, 1hp = 745.700W). This unit was defined in the 18th century by James Watt. Watt sold steam engines and was asked how many horses one steam engine would replace. He had horses walk around a wheel that would lift a weight. He found that each horse would average about 550 foot-pounds of work per second. One horsepower is equivalent to 550 foot-pounds per second [550(ft-lb)/s] or 33 000 foot-pounds per minute [33 000(ft-lb)/min].

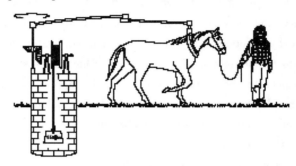

The following formula can be used to calculate horsepower when torque (M) (in lb-ft) and speed (RPM) are known. An increase of torque, speed, or both will cause an increase in horsepower.

$$HP = \frac{M \times RPM}{5\,250}$$

New Words and Phrases

AC (Alternating Current)		交流电
AC motor		交流电动机
residential	[ˌreziˈdenʃəl]	adj. 住宅的,民用的
commercial	[kəˈməːʃəl]	adj. 商业的,商用的
utility	[juːˈtiliti]	n. 公用事业
electrical energy		电能
mechanical energy		机械能
mechanical equipment		机械设备
pump	[pʌmp]	n. 泵,抽水机
		vt. (用泵)抽(水),抽吸

winder	[ˈwaɪndə]	n. 卷扬机
conveyor	[kʌnˈveɪə(r)]	n. 传送装置,传送带
mixer	[ˈmɪksə]	n. 搅拌器,搅拌机;混频器
terminology	[ˌtɜːmɪˈnɒlədʒɪ]	n. 术语学
associate with		与……相关,同……联系在一起
in simple term		简言之
electromagnetism	[ˌɪlektrəʊˈmægnɪtɪz(ə)m]	n. 电磁,电磁学
net force		净力
vector	[ˈvektə]	n. 向量,矢量
object	[ˈɒbdʒɪkt]	n. 物体,物品,实物;对象
opposite	[ˈɒpəzɪt]	adj. 相反的,相对的,对面的,对立的
friction	[ˈfrɪkʃən]	n. 摩擦,摩擦力
torque	[tɔːk]	n. 扭矩,转矩
twisting	[ˈtwɪstɪŋ]	n. 扭曲
rotate	[rəʊˈteɪt]	vt. & vi. (使)旋转
pivot	[ˈpɪvət]	n. 枢轴,支点
product	[ˈprɒdəkt]	n. 乘积
law of inertia		惯性定律
radius	[ˈreɪdjəs]	n. 半径,范围,界限
measure	[ˈmeʒə]	n. 测量,量度器,量度标准 vt. 测量,测度,估量,调节
corresponding	[ˌkɒrɪˈspɒndɪŋ]	adj. 相应的
travel	[ˈtrævl]	n. 行程,冲程,动程
linear	[ˈlɪnɪə]	adj. 线的,直线的,线性的
distance	[ˈdɪstəns]	n. 距离,间隔
meters per second		米每秒
lever	[ˈliːvə, ˈlevə]	n. 电平;杆,杠杆,控制杆 vt. & vi. 抬起
ratio	[ˈreɪʃɪəʊ]	n. 比,比率
angular	[ˈæŋgjulə]	adj. 有角的
angular speed		角速度
revolution	[ˌrevəˈluːʃən]	n. 旋转
RPM (Revolutions Per Minute)		转/分(r/min)
acceleration	[ækˌseləˈreɪʃən]	n. 加速度
deceleration	[diːˌseləˈreɪʃən]	n. 减速
inertia	[ɪˈnɜːʃjə]	n. 惯性,惯量
external force		外力

moment of inertia		n.	惯性(力)矩,转动惯量
property	['prɔpəti]	n.	性质,特性
overcome	[,əuvə'kʌm]	vt.	克服
compensate	['kɔmpənseit]	vt. & vi.	补偿,偿还
loss	[lɔs]	n.	损耗
bearing	['bɛəriŋ]	n.	轴承
work	[wə:k]	n.	功
accomplished	[ə'kɔmpliʃt]	adj.	完成的
define	[di'fain]	vt.	定义,详细说明
express	[ik'spres]	vt.	表达,表示
steam engine			蒸汽机
equivalent	[i'kwivələnt]	adj.	相等的,相当的
		n.	等价物,相等物
horsepower	['hɔ:s,pauə]	n.	马力
hp (horsepower)			马力
walk around…			绕……而走

Notes

[1] AC motors are found on a variety of applications from those that require a single motor to applications requiring several motors.

本句的基本结构是：主语 AC motors,谓语：are found,from…to…是 application 的范围。其中 that require a single motor 是定语从句,修饰 those,requiring several motors 是一个现在分词短语修饰它前面的 applications。

[2] The angular speed of a rotating object is a measurement of how long it takes for a given point on the object to make one complete revolution from its starting point.

本句可以分成几个层次理解。第一层是说角速度是一种量度(the angular speed is a measurement),第二层说的是角速度量度什么(a measurement of how long it takes),第三层是角速度如何量度(a given point make(s) one complete revolution from its starting point,物体从给定起始点完成一周旋转)。

[3] If we look at a continuous roll of paper, for example, we know that when the roll is stopped it would take a certain amount of force to overcome the inertia of the roll to get it rolling.

本句是关于 roll 的理解。roll 既可作动词(意为卷动)也可作名词(意为一卷)。很显然本句中的 roll 是名词,a continuous roll of paper 指一卷纸,when the roll is stopped 以及 the inertia of the roll 中的 roll 都是指卷纸。而最后的 rolling 是个现在分词,表示"滚动"。stopped 不是"静止"而是"停止"。

[4] Once the system is in motion only the energy required to compensate for various losses need be applied to keep the conveyor in motion.

本句是包含后置定语的复合句。once 表明条件：一旦……。主句中 required to compensate for various losses 是过去分词短语修饰主语 energy,动词不定式短语 to keep the conveyor in motion 作为 applied 的目的。

[5] Friction within motor and driven equipment bearings; Friction between material on winder and rollers.

这是由相同中心词 friction(摩擦)组成的两个名词短语,因为介词不同,所表达意思也不同。第一个 friction 后用的是 within,意思为在……之内,即"电动机和驱动设备轴承中的摩擦"。第二个 friction 后用的是 between…and…,指的是两者之间的摩擦,即物体和"卷扬机、滚筒之间的摩擦"。

Exercises

【Ex.1】 根据课文内容,回答以下问题。

(1) What application can AC motors be used in?

(2) What is NET force?

(3) What is the content of the law of inertia?

(4) What is power?

(5) What is one horsepower equivalent to?

【Ex.2】 根据下面的英文解释,写出相应的英文词汇。

英 文 解 释	词　汇
something, such as a machine or an engine, that produces or imparts motion	
a machine or device for raising, compressing, or transferring fluids	
the natural force of attraction exerted by a celestial body, such as earth, upon objects at or near its surface, tending to draw them toward the center of the body	
the rubbing of one object or surface against another	
a line segment that joins the center of a circle with any point on its circumference	
to increase the speed of	
the tendency of a body to resist acceleration; the tendency of a body at rest to remain at rest or of a body in motion to stay in motion in a straight line unless acted on by an outside force	
a unit of power in the U.S. Customary System, equal to 735.5 W or 33 000ft-lb per minute	

英 文 解 释	词 汇
the capacity for work or vigorous activity; vigor; power	
a solid disk or a rigid circular ring connected by spokes to a hub, designed to turn around an axle passed through the center	
to turn around on an axis or center	
the moment of a force; the measure of a force's tendency to produce torsion and rotation about an axis, equal to the vector product of the radius vector from the axis of rotation to the point of application of the force	

【Ex. 3】 把下列句子翻译为中文。

(1) The magnetic moment can be considered to be a vector quantity with direction perpendicular to the current loop in the right-hand-rule direction.

(2) Magnetic fields are produced by electric currents, which can be macroscopic currents in wires, or microscopic currents associated with electrons in atomic orbits.

(3) Magnetic field sources are essentially dipolar in nature, having a north and a south magnetic poles.

(4) The electric force is straightforward, being in the direction of the electric field if the charge Q is positive, but the direction of the magnetic part of the force is given by the right-hand-rule.

(5) For circular motion at a constant speed v, the centripetal acceleration of the motion can be derived.

(6) For an object rotating about an axis, every point on the object has the same angular velocity. The tangential velocity of any point is proportional to its distance from the axis of rotation.

(7) These rotation equations apply only in the case of constant angular acceleration. It is assumed that the angle is zero at $t=0$ and that the motion is being examined at time t.

(8) Moment of inertia is the name given to rotational inertia. It appears in the relationships for the dynamics of rotational motion.

(9) The force arm is defined as the perpendicular distance from the axis of rotation to the line of action of the force.

(10) It is calculated as the product of the force and the distance through which the body moves and is expressed in joules(J), ergs(erg), and ft-lb.

【Ex. 4】 把下列短文翻译成中文。

If the motor runs at 600r/m when the PWM drive to the motor is at a 50% duty ratio, increasing this to 60% will make the motor try to run faster. Reducing the duty ratio to 40% will slow the motor down. In a very simple "open loop" speed controller, the potentiometer on analog channel 0 is read to yield a value between 0 ~ 1 023 (10bit). This value is then fed into the PWM unit to allow the motor speed to be varied.

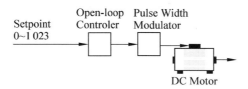

In the real world, this type of controller is not very useful. While it allows the motor speed to be set, it does not allow for changes in load and a basic flaw is that the absolute speed is not known unless an external tachometer is used.

【Ex. 5】 通过Internet查找资料,借助如"金山词霸"等电子词典和辅助翻译软件,完成以下技术报告。通过E-mail发送给老师,并附上你收集资料的网址。

(1) 当前世界上有哪些最主要的电动机生产厂家以及有哪些最新型的产品(附上各种最新产品的图片)。

(2) 通过Internet查找关于电动机控制的基本原理和电路。

Text B Basic DC Motor Operation

1. Magnetic Fields

You will recall from the previous section that there are two electrical elements of a DC motor, the field windings and the armature. The armature windings are made up of current carrying conductors that terminate at a commutator. DC voltage is applied to the armature windings through carbon brushes which ride on the commutator. In small DC motors, permanent magnets can be used for the stator. However, in large motors used in industrial applications the stator is an electromagnet. When voltage is applied to stator windings an electromagnet with north and south poles is established. The resultant magnetic field is static (nonrotational). For simplicity of explanation, the stator will be represented by permanent magnets in the following illustrations.

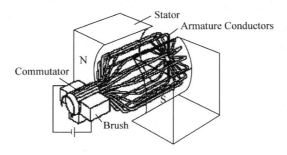

2. Magnetic Fields

A DC motor rotates as a result of two magnetic fields interacting with each other. The first field is the main field that exists in the stator windings. The second field exists in the armature. Whenever current flows through a conductor a magnetic field is generated around the conductor.

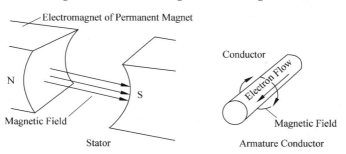

3. Right-Hand Rule for Motors

A relationship, known as the right-hand rule for motors, exists between the main field, the field around a conductor, and the direction the conductor tends to move. If the thumb, index finger, and third finger are held at right angles to each other and placed as shown in the following illustration so that the index finger points in the direction of the main field flux and the third finger

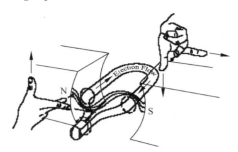

points in the direction of electron flow in the conductor, the thumb will indicate direction of conductor motion. As can be seen from the following illustration, conductors on the left side tend to be pushed up. Conductors on the right side tend to be pushed down. This results in a motor that is rotating in a clockwise direction. You will see later that the amount of force acting on the conductor to produce rotation is directly proportional to the field strength and the amount of current flowing in the conductor.

4. CEMF

Whenever a conductor cuts through lines of flux a voltage is induced in the conductor. In a DC motor the armature conductors cut through the lines of flux of the main field. The voltage induced into the armature conductors is always in opposition to the applied DC voltage. Since the voltage induced into the conductor is in opposition to the applied voltage it is known as CEMF (Counter Electromotive Force). CEMF reduces the applied armature voltage.

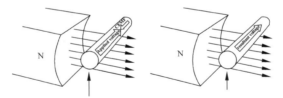

The amount of induced CEMF depends on many factors such as the number of turns in the coils, flux density, and the speed which the flux lines are cut.

5. Armature Field

An armature, as we have learned, is made up of many coils and conductors. The magnetic fields of these conductors combine to form a resultant armature field with a north and a south poles. The north pole of the armature is attracted to the south pole of the main field. The south pole of the armature is attracted to the north pole of the main field. This attraction exerts a continuous torque on the armature. Even though the armature is continuously moving, the resultant field appears to be fixed. This is due to commutation, which will be discussed next.

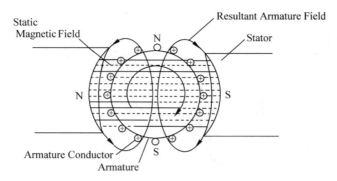

6. Commutation

In the following illustration of a DC motor only one armature conductor is shown. Half of the conductor has been shaded black, the other half white. The conductor is connected to two segments of the commutator. In position 1 the black half of the conductor is in contact with the negative side of the DC applied voltage. Current flows away from the commutator on the black half of the conductor and returns to the positive side, flowing towards the commutator on the white half.

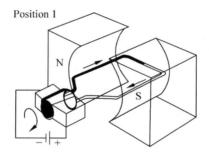

In position 2 the conductor has rotated 90°. At this position the conductor is lined up with the main field. This conductor is no longer cutting main field magnetic lines of flux; therefore, no voltage is being induced into the conductor. Only applied voltage is present. The conductor coil is short-circuited by the brush spanning the two adjacent commutator segments. This allows current to reverse as the black commutator segment makes contact with the positive side of the applied DC voltage and the white commutator segment makes contact with the negative side of the applied DC voltage.

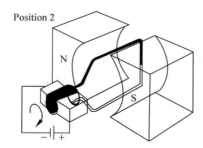

As the conductor continues to rotate from position 2 to position 3 current flows away from the commutator in the white half and toward the commutator in the black half. Current has reversed direction in the conductor. This is known as commutation.

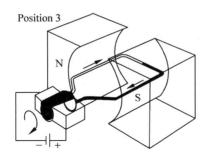

New Words and Phrases

DC motor			直流电动机
winding	['waindiŋ]	n.	绕,缠,绕组,线圈
terminate	['tə:mineit]	vt. & vi.	停止,结束,终止
commutator	['kɔmjuteitə]	n.	换向器,转接器
carbon brush		n.	碳刷
magnet	['mægnit]	n.	磁体,磁铁
stator	['steitə]	n.	定子,固定片
electromagnet	[ilektrəu'mægnit]	n.	电磁石
nonrotational	['nɔnrəu'teiʃənl]	adj.	无旋的
permanent magnet			永久磁铁
right-hand rule			右手定则
thumb	[θʌm]	n.	拇指
index finger			食指
third finger			中指
CEMF (Counter Electro Motive Force)			反电动势
coil	[kɔil]	n.	线圈;卷,圈,盘卷,圈形物;一卷,一圈
flux	[flʌks]	n.	流量,通量
		vi.	熔化,流出
		vt.	使熔融
armature field			电枢场
resultant	[ri'zʌltənt]	adj.	作为结果而发生的,合成的
south pole			南极
north pole			北极
magnetic lines			磁力线

Exercises

【Ex. 6】 根据文章所提供的信息判断正误。

(1) There are two electrical elements of a DC motor, the field windings and the stator.

(2) In all DC motors, permanent magnets can be used for the stator.

(3) Because two magnetic fields interact with each other, a DC motor works.

(4) If you know the direction of electron flow in the conductor and the direction of the main field flux, you can decide the direction of conductor motion according to the right-hand rule.

(5) If a conductor cuts through lines of flux, the voltage in the conductor is caused.

(6) Whenever a conductor cuts through lines of flux a voltage is induced in the conductor. The voltage is CEMF.

(7) The north pole of the armature is attracted to the south pole of the main field. The south pole of the armature is attracted to the north pole of the main field.

(8) When the armature is moving, the resultant armature field will be changed.

(9) The commutator can keep armature field to be fixed when armature is moving.

(10) The commutator can change current direction in the conductor.

科技英语翻译知识　否定的译法

在英汉翻译,特别是科技英语翻译中,否定含义的理解和表达不容忽视。英语里表示否定意义的词汇、语法手段与汉语有很大不同。英语里的否定句有全部否定与部分否定之分;也有字面结构为肯定句,但实际含有否定意义的;还有实为强烈肯定的双重否定。翻译中必须正确理解才能正确表达。

1. 全部否定

Not, no, never, none, neither...nor, nothing, nowhere 等否定意义强烈的词出现在句中时,句子表示完全否定。翻译这类句子时,把表示否定的"不、无、非"等的词与动词连用,即可达到全部否定。

(1) Rubber does not conduct electricity.

橡胶不导电。

(2) None of these metals have conductivity higher than copper.

这些金属中没有一种其电导率比铜高。

(3) A proton has a positive charge and an electron has a negative charge, but a neutron has neither.

质子带正电荷,电子带负电荷,但中子不带任何电荷。

(4) A semiconductor is a material that is neither a good conductor nor a good insulator.

半导体材料既不是良导体,也不是好的绝缘体。

(5) There is no steel not containing carbon.

没有不含碳的钢。

2. 部分否定

由 all, every, both, always 等含有全体意义的词与否定词 not 构成部分否定。翻译时要特别注意,因为这样的搭配看似全部否定,而实际却是部分否定。

(1) All these metals are not good conductors.

这些金属并不都是良导体。

(2) Not all substances are good conductors of electricity.

并非所有的物质都是电的良导体。

(3) Both of the instruments are not precise.

这两台仪器不都是精密的。

(4) Every machine here is not imported from abroad.

这里的机器并非每台都是从国外进口的。

(5) But friction is not always useless, in certain cases it becomes a helpful necessity.

可是摩擦并不总是没用的,在某些情况下是有益而且必需的。

3. 意义否定

有些英语句子,在结构上是肯定句,因为包含带否定意义的词或词组,实际却是否定句。

(1) Slight amount of current lowing through the voltmeter may be ignored.

通过电压表的微量电流是可以忽略不计的。

(2) If the lighting or supply to other services fails, the battery supply takes over.

假如照明或别的供电失灵,蓄电池将承担供电任务。

(3) The purpose of the fuse is to protect such equipment from excessive current.

保险丝的用途是保护上述设备不受过量电流的影响。

(4) Another advantage of the absence of moving parts is that a transformer needs very little attention.

变压器没有运转部件的另一好处在于它几乎不需要关注。

(5) The atom left deficient by losing one unit of negative charge and becomes a positive ion.

该原子失去一个单位的负电荷而变为正离子。

4. 双重否定

两个否定意义的词连用构成双重否定。这种否定实际是语气强烈的肯定,翻译时既可译成双重否定,也可转译成肯定句。

(1) No machines in the modern factories can work without electricity.

在现代工厂中,机器没有电就不能工作。

(2) Indeed, it is possible that electrons are made of nothing else but negative electricity.

的确,电子只可能是由负电荷构成的。

(3) A radar screen is not unlike television screen.

雷达荧光屏跟电视荧光屏没什么不一样。

(4) There has not been a scientist of eminence but was a man of industry.

没有哪一个有成就的科学家不是勤奋的。

(5) There is no rule that has no exceptions.

没有无例外的规则。

Reading Material

阅读下列文章。

Text	Note
Protel DXP Specifications **Schematic Capture Specifications**[1] • Max Sheet Size: 64×64 inches • Max. Sheet Resolution: 0.1 inches • Max. Sheets Per Project: Unlimited[2] • Sheet hierarchy: Unlimited depth • Font support: All Windows-supported fonts • Output Device Support: All Windows output devices • Netlist Output Formats: Protel; EDIF 2.0 for PCB; EDIF 2.0 for FPGA; CUPL PLD; MultiWire; Spice3f5; VHDL • Max. Components Per Sheet: Unlimited • Max. Pins[3] Per Component: Unlimited • Max. Components Per Library: Unlimited • Max. Simultaneously[4] Open Libraries: Unlimited • Electrical Drawing Tools: Bus; Bus Entry; Component Part; Junction; Power Port; Wire; Net Label; Port; Sheet Symbol; Sheet Entry • Non-Electrical Drawing tools: Text Annotation[5]; Text Box; Arc; Elliptical Arc; Ellipse; Pie Chart; Line; Rectangle; Rounded Rectangle; Polygon; 4-point Bezier[6]; Graphic Image • Assignable Component Pin Electrical Types: Input; Output; Input/Output; Open Collector; Passive; Hi Z; Emitter[7]; Power; VHDL Buffer[8]; VHDL Port • User-definable Component Parameters: unlimited, at both the library editor and schematic sheet • Report Generation: Bill Of Materials; Project Hierarchy[9]; Cross Reference • Import File Formats: all Protel schematic[10] formats; AutoCAD DXF/DWG up to 2000; P-CAD Schematic ASCII (V15 & V16); Orcad Capture (V7 & V9) • Export File Formats: Orcad DOS schematic; Protel schematic V4; Protel ASCII; Protel schematic template ASCII and binary	[1] n. 规范,说明书,详述,规格 [2] adj. 不受限制的 [3] n. 插脚 [4] adv. 同时地 [5] n. 注解,注释 [6] n. 贝塞尔曲线 [7] n. 发射体,发射极 [8] n. 缓冲器 [9] n. 分层,层级 [10] adj. 示意性的

PCB Layout Editor Specifications

- 74 Layers: 32 × Signal; 16 × Mechanical; 16 × Internal Plane; 2 × Solder Mask; 2 × Paste Mask; 2 × Silkscreen; 2 × Drill (Drill Guide & Drill Drawing); 1 × Keep Out; 1 × Multi-Layer (spans all signal layers)
- Max. board size: 100 × 100 inch
- Max. Resolution[11]: 0.001 mil[12] linear displacement; 0.001 degree angular rotation
- Import File Formats: Netlist[13] (Protel and Tango); all Protel PCB formats; AutoCAD DXF/DWG up to R14; Gerber-batch or single; P-CAD PCB ASCII (V15 & V16); PADS PCB ASCII (various up to 3.5); Orcad Layout (V7); Specctra RTE
- Export File Formats: AutoCAD DXF/DWG; Specctra DSN; HyperLynx; Protel Netlist; Protel V2.8 ASCII, V3 binary, V4 binary, V5 ASCII
- Output Device Support: Gerber RS274X (internal driver, supports embedded apertures[14]); all Windows printer & plotter drivers
- Measurement Units: Metric and Imperial
- Max. Net Count: Unlimited
- Report Generation: Board Information Summary; Bill Of Materials; Design Hierarchy; Signal Integrity; NC Drill (Excellon binary and ASCII); Pick and Place; Create Netlist from copper
- Pads Styles: Round; Rectangular; Octagonal[15]
- Pad Size: 0.001 to 99 999 mil
- Pad Stacks[16]: Simple (all layers the same); Top-Mid-Bottom (different shapes definable for top, bottom and mid layers); Full Stack (each layer defined independently)
- Via Styles: Through-hole; Blind & Buried (Layer Pairs); Blind & Buried (Any Layers)
- Track Widths: 0.001 to 99 999 mil
- Track Placement Modes: Oblique[17]; 45 deg; 45 deg with arcs; 90 deg; 90 deg with arcs
- Interactive Routing Modes: Ignore Obstacles[18]; Avoid Obstacles; Push-and-Shove
- Polygon Plane Styles: 90 deg hatched[19]; 45 deg hatched; Vertical hatch; Horizontal hatch; Solid
- Plane Connectivity: Assignable to any net
- Power Planes: All plane layers can be assigned to any net, and all plane layers can have multiple splits[20]
- Max. Components Per Board: Unlimited
- Max. Pins Per Component: Unlimited
- Max. Components Per Library: Unlimited
- Max. Simultaneously Open Libraries: Unlimited

Autorouting[21] Specifications

- Routing method: Topological[22] mapping

[11] n. 分辨率
[12] n. 千分之一英寸
[13] n. 网表
[14] n. 孔,孔径
[15] adj. 八边形的,八角形的
[16] n. 堆栈,堆叠
[17] adj. 倾斜的,间接的
[18] n. 障碍
[19] adj. 阴影线的
[20] n. 缝
[21] n. 自动布线
[22] adj. 拓扑的

• Routing passes: Memory; Fan out; Pattern; Push-&-Shove; Rip up; Track spacing; Testpoint addition • Max. Components: unlimited • Max. Pins/Component: 5 000 • Max. Nets: 10 000 • Max. Connections: 16 000 **Signal Integrity Specifications** • Device modeling: I/O buffer macro-model • Analyses: Net Screening[23]; Reflection; Crosstalk[24] • Integration methods: Trapezoidal; Gear's Method (1st, 2nd and 3rd order) • Termination simulations: Series R; Parallel R to V_{CC}; Parallel R to GND; Parallel R to V_{CC} and GND; Parallel C to V_{CC}; Parallel C to GND; Parallel diodes • Stimulus[25] types: Single pulse[26]; Constant level; Periodic pulses **Mixed Signal Circuit Simulation Specifications** • Analyses: Operating point; Transient[27]; Fourier; AC small signal; DC sweep[28] (2 variables); Noise[29]; Transfer function; Temperature sweep (2 variables); Parameter sweep (2 variables); Monte Carlo • Analog modeling: Spice 3f5 compatible • Digital modeling: Digital SimCode (XSpice compatible) • AC sweep types: Linear; Decade; Octave • Monte Carlo distributions: Uniform; Gaussian[30]; Worst Case **Programmable Logic Design Specifications** • Download[31] formats: JEDEC; POF; PRG; HL (Signetics IFL devices); ASCII hex • Output formats: Palasm PDS; Expanded Macro MX; Berkely PLA; Xilinx XNF; PDIF; EDIF • Minimization methods: Quick; Presto[32]; Expresso **CAM Output** • Import formats: QuickLoad (load all supported file types simultaneously); ODB++; Netlists (IPC-D356); Gerber data (274D, 274X, Fire9000); HPGL/2, NC Drill & Mill/Route; DWG/DXF (AutoCAD®); Any Aperture file (using aperture Wizard[33]); Aperture Libraries (*.LIB); CAMtastic® database file (*.CAM) • Export[34] formats: ODB++; Netlists (IPC-D356); IPC-D350; Gerber data (274D, 274X, Fire9000); NC Drill & Mill/Route; DXF (compatible with any CAD system); Aperture Libraries (*.LIB); CAMtastic® database file (*.CAM) • Verification and analysis tools: PCB Design Check/Fix (18 different manufacturing checks); Compare Netlists; Netlist Extraction (supports blind/buried vias); Measure (Object to Object, Point to Point); Invalid Polygon[35] Search; Calculate Copper Area and Compare Layers • NC drill and route tools: Create/Modify Drill & Rout paths; Add Drill	[23] *n.* 筛选 [24] *n.* 串扰 [25] *n.* 激励 [26] *n.* 脉冲 [27] *adj.* 瞬时的 [28] *vt. & vi.* 扫频 [29] *n.* 噪声 [30] *adj.* 高斯的 [31] *vt. & vi.* 下载, 下传 [32] *adj.* 极快的, 迅速的 [33] *n.* 向导 [34] *vt. & vi.* 导出 [35] *n.* 多角形, 多边形

(Point, Circle, Slot, Text); Build Drill layer (Plated Thru Hole); Auto-Rout PCB Border; Convert segments to Arcs/Circles; Add Tabs[36] and Mill Boundary	[36] *n.* 制表符
• Editing and modification tools: Panelize PCB; Group Objects; Film Wizard; Pad Removal; Copper Pour; Teardrops; Spread/Reduce; Trim Silkscreen[37]; Venting; Clean Boundaries; Generate Outlines; Step/Repeat; Rotate, Mirror, Scale (X:Y), Join; Align Layers; Create Composites and Explode	[37] *n.* 丝网印刷
• Place features: Flash; Text; Line; Polyline; Rectangle; Ellipse; Polygon; Polygon Void[38]; Arc; Circle; Dimensions	[38] *adj.* 空的,无效的,无用的
• Reporting features: PCB RFQ (Request for Quotation); Drill, Dcode/Layer Usage; DRC/DFM; Netlist; Rout/Mill; Status and X:Y Coordinates	
• Print features: Complete Windows printing support; Print Area; Print to Scale; Print Preview; Separate page for each layer; Group multiple images on each page	
FPGA Synthesis Multi-vendor device support, including: • Xilinx, 3k, 4k, 5k, 7k, 9.5k, Spartan, Virtex and Coolrunner Series • Altera Stratix, APEX, Cyclone, ACEX, FLEX, MAX and Excalibur Series • Actel ACT, 40MX, 42MX, 54SX, eX, 500k and proAsic Series • Atmel PLD • Lattice PLSI and ORCA Series • Quicklogic PASIC Series • Vantis CPLD	

参考译文 交流电动机

交流电动机被广泛地使用在民用、商业、工业和公共事业项目中。电动机将电能转化为机械能。一个交流电动机可以是泵或风扇的一部分,也可以连接到卷扬机、传送机或搅拌机等其他机械设备中。交流电动机可以被广泛应用到需要一台或多台电动机的场合。

1. 力和运动

在讨论交流电动机之前,有必要先了解一些与电动机运行相关的术语。许多术语我们已经在别的文章里熟悉了,一会儿我们要了解这些术语在交流电动机上的应用。

2. 力

简言之,力就是推或拉。力可以是由电磁力、重力或一些物理方法组合形成。

3. 净力

净力是作用于一个物体的所有力的向量和,包括摩擦力和重力。当这些力从相同方向施加时,则它们是相加的。例如,两个 10N 的力从相同的方向作用,则净力是 20N。

如果一个 10N 的力从一个方向作用,另一个 20N 的力从一个相反的方向作用,则净力是 10N,物体将沿大的力的方向移动。

如果两个 10N 的力,以相反的方向作用某物体,则净力为零,物体静止不动。

4. 转矩

转矩是引起物体旋转的扭曲力或转动力。例如,在杠杆的一端施加一个力,对于支点会形成旋转作用或力矩。力矩(M)是力和半径(杠杆的长度)的乘积。

$$M = 力 \times 半径$$

在英制中力矩是用磅-英尺(lb-ft)或磅-英寸(lb-in)来计量的。例如,10N 的力作用到 1ft 长的杠杆上,将产生 10lb-ft 的力矩。

增加力或半径将会相应增加力矩。例如,将半径增加到 2ft,结果将产生 20lb-ft 的力矩。

5. 速度

运动物体在给定的时间移动一段距离。速度是距离与通过这段距离所用时间之比。

$$速度 = 距离 / 时间$$

6. 线速度

物体线速度是从 A 点到 B 点快慢度量。线速度一般是以米每秒(m/s)的形式表示。例如,如果 A 和 B 两点的距离是 10m,移动这段距离要用 2s,速度就是 5m/s。

7. 角速度

旋转物体的角速度是物体从给定起始点旋转一周所用时间的量度。角速度的单位是弧度/分。例如,物体在 1min 完成 10 周旋转具有的角速度是 62.8rad/min。

8. 加速度

物体可以做变速运动。速度的增加量称为加速度。只有在物体上有作用力时才会产生加速度。物体也能从高速变为低速,这就是减速(负加速)。例如,旋转物体能够从 62.8rad/min 加速到 125.6rad/min,或从 125.6rad/min 减速到 62.8rad/min。

9. 惯性

机械系统服从惯性定律。惯性定律陈述的是:物体在受到外力的作用之前保持当前的静止或运动状态。抵抗加速或减速的属性就是人们所说的惯性。英制的计量系统以磅-英尺^2(lb-ft^2)为单位。例如一卷纸,我们知道当卷纸在停止时,必须用一定的力才能克服卷纸的惯性使它滚动。这种克服惯性所需要的力可以来自某些能量源,比如电动机。一旦滚动起来,卷纸将不断滚动,直到另外一个力作用于它,才能停止。

10. 摩擦

力的大部分被用于克服静止系统的惯性以启动物体移动。由于摩擦消耗了机械系统的能量,为了维持物体的运动必须不断施加力。但是惯性定律始终有效,因为施加的力只需要补偿能量的损耗。一旦系统已经运动,要维持设备的运转,只需要补偿各种能量损耗。例如,在前面图例中,这些损耗包括:
- 电动机内部和驱动设备轴承的摩擦;
- 电动机以及驱动设备的转动损耗;
- 卷扬机和滚筒之间的摩擦。

11. 功

只要任何一种力作用引起物体运动,该力就做了功。例如,运输机上的物体从一个点被移动到另一个点,就做了功。

功一般是用英尺-磅(ft-lb)表示,是施加的净力(F)与移动的距离(d)的乘积。如果施加二倍的作用力,所做的功也是原来的二倍。如果移动的距离是二倍,所做的功也是原来的二倍。

$$W = F \times d$$

12. 功率

功率是做功的效率,或做的功除以所用的时间。

$$功率 = (力 \times 距离)/时间$$

换句话说,功率就是将一个物体从一个点移动到另一个点所做功的总和除以所用的时间。

13. 马力

功率可以用英尺-磅每秒(ft-lb)/s 表示,但常常是用马力(hp)来表示。这个单位是由 18 世纪的詹姆士·瓦特定义的。瓦特在卖蒸汽机时,有人问一台蒸汽机能代替多少匹马。他让马匹绕着一个可以提升重物的轮子行走。他发现平均每匹马能在 1s 内做 550 英尺-磅的功。1 马力相当于每秒做 550 英尺-磅[550(ft-lb)/s]或每分做 33 000 英尺-磅的功(即 33 000(ft-lb)/min)。

下面的公式可被用于计算速度和力矩给定时的马力(hp)。分别增加力矩(M)、速度(RPM,转每分)或同时增加力矩和速度都会增加马力。

$$hp = \frac{M \times RPM}{5\ 250}$$

Unit 7

Text A The Basis of Control

1. The principle of closed-loop control

The principle of closed-loop control consists in the value to be controlled being fed back from the place of measurement via the controller and its settings' facility into the controlled system. [1] The feedback process makes this so-called controlled variable more independent of external and internal disturbance variables, and is the factor which enables a desired value, the setpoint, to be adhered to in the first place. [2] As the manipulated variable output by the controller influences the controlled variable, the so-called control loop is duly closed.

Technical systems process several kinds of controlled variables such as current, voltage, temperature, pressure, flow, speed of rotation, angle of rotation, chemical concentration and many more. Disturbance variables are also of a physical nature.

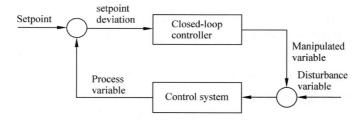

The principle of closed-loop control

These control loop terms can be easily explained using the familiar example of room temperature control by means of a radiator thermostat: the requirement is to keep the room temperature at 22℃. This temperature is set by means of the rotatable knob (= setpoint). The temperature (= controlled variable) is measured by a sensor. The deviation between the room temperature and the setpoint is then measured by the built-in controller, often a bimetal spring, (= the control deviation), and is then used to open or close the valve (= the manipulated variable).

What are the disturbance variables? First of all there is the effect of the outside temperature and the sun shining through the windows. The "limited" thermostat can as little foresee these influences as it can the occupant's behaviour in opening a window or holding a party with a lot of guests who cause the room to warm up. [3] However this controller is still able to compensate for the effects of one or more disturbance variables, and to bring the temperature back to the desired level again, albeit with some delay.

2. The principle of open-loop control

Open-loop control is found wherever there is no closed-control loop. The biggest disadvantage compared with closed-loop control, is that unknown or non-measurable disturbance variables cannot be compensated. Also the behaviour of the system including the effects of disturbance variables which the open-loop control system is able to measure, must be exactly known at all times in order to be able to use the manipulated variable to influence the controlled variable. [4]

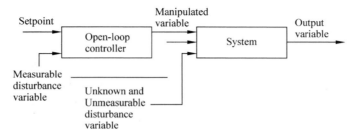

The principle of open-loop control

One advantage is that an open-loop control system cannot become instable as there is no feedback — this is a problem of closed-loop control.

3. Classification of closed-loop systems

Closed-loop systems are not classified by the physical values to be controlled, but by their behaviour over time. The level in a container can thus be mathematically described in exactly the same way as the voltage of a capacitor.

Behaviour over time can be determined, for example, by abruptly changing the input value and then observing the output value. Knowledge of basic laws of physics is often sufficient to estimate this behaviour. Only in relatively few cases is necessary to calculate.

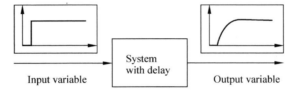

Control system with delay

The behaviour over time of a closed-loop system is normally characterised by the fact that when the input value is abruptly changed, although the output value immediately begins to change, it reaches its end value with some delay. [5]

Closed-loop systems are further distinguished by those with and those without self-regulation.

In a system with self-regulation, after a sudden change in the input value, the output value resume a constant value again after a period of time. Such systems are usually called proportional systems or P systems. Let us take the example of a heating zone: the input value is the electrical heating power, and the output value is the zone temperature.

Control system with self-regulation

In a system which does not have self-regulation, the output value will rise or fall after the abrupt change in the input value. The output will only remain at a constant value if the input value is at zero. Such systems are usually called integral systems or I systems. An example of this is a level control in a container: the input value is the incoming flow, the output value is the level of the liquid.

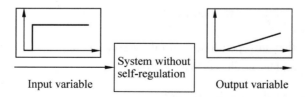

Control system without self-regulation

Another important type of system is system with dead time. In this case the input value appears at the output after the dead time delay. In a technical system the dead time is the result of the distance between setting and measuring locations. Example of a conveyor belt: the input value is the quantity of material at the beginning of the belt, and the output value is the measurement of the amount at the end of the belt. The dead time is calculated as the length of the belt divided by its speed, and it can therefore vary.

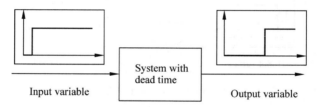

Control system with dead time

In all the different systems discussed which are normally found in combination, we are dealing with so-called linear single variable systems because there is only one output value (= the controlled variable) as well as one input value (= the manipulated variable), and the system possesses linear properties.[6] Controllers for these very common systems are accordingly known as single-variable control systems.

New Words and Phrases

principle	['prinsəpl]	n.	法则,原则,原理
closed-loop		n.	闭环
consist in			存在于
controller	[kən'trəulə]	n.	控制器
facility	[fə'siliti]	n.	设备,工具
feedback	['fi:dbæk]	n.	反馈,反应
independent of			不依赖……,独立于……
factor	['fæktə]	n.	因素,要素
duly	['dju:li]	adv.	适当地,合适地,适度地;
adhere to			坚持
built-in		adj.	内置的,固定的,嵌入的
disturbance	[dis'tə:bəns]	n.	打扰,干扰
compare with…			与……比较
manipulate	[mə'nipjuleit]	vt.	(熟练地)操作,使用(机器等)
thermostat	['θə:məstæt]	n.	自动调温器,温度调节装置
rotatable	['rəuteitəbl]	adj.	可旋转的,可转动的
knob	[nɔb]	n.	(门,抽屉等的)球形把手,节,旋钮
sensor	['sensə]	n.	传感器
deviation	[,di:vi'eiʃən]	n.	偏离,偏向,偏差
bimetal	[bai'metl]	n.	双金属材料,双金属器件
		adj.	双金属的(= bimetallic)
spring	[spriŋ]	n.	弹簧,发条;弹性,弹力
warm up		vt. & vi.	变暖,热身
disadvantage	[,disəd'va:ntidʒ]	n.	缺点,不利,不利条件,劣势
classify	['klæsifai]	vt.	分类,分等
over time		n.	超时,滞后
behaviour	[bi'heivjə]	n.	行为,举止,习性
sufficient	[sə'fiʃənt]	adj.	充分的,足够的
abruptly	[ə'brʌptli]	adv.	突然地,唐突地
relatively	['relətivli]	adv.	相关地
characterise	['kærəktəraiz]	vt.	表现……特点,具有……特

		征(= characterize)
estimate	['estimeit]	vt. & vi. 估计,估价,评估
		n. 估计,估价,评估
proportional system		比例系统
self-regulation		自调整
constant	['kɔnstənt]	n. 常数,恒量
		adj. 不变的,持续的
dead time		死区
belt	[belt]	n. 带子

Notes

[1] The principle of closed-loop control consists in the value to be controlled being fed back from the place of measurement via the controller and its settings facility into the controlled system.

本句是个简单句,谓语动词是 consists in(存在于……),宾语由两部分组成:value to be controlled being fed back from the place of measurement via the controller 以及 its settings' facility into the controlled system。这两个宾语都是由名词短语组成,分别表示两个动作:被控制的值反馈回来(value being fed back)以及设定的值迅速进入控制系统(settings' facility into the controlled system)。

[2] The feedback process makes this so-called controlled variable more independent of external and internal disturbance variables, and is the factor which enables a desired value, the setpoint, to be adhered to in the first place.

这句话是一个由 and 连接的并列句,两个并列句有一个共同的主语 feedback process。第一个句子的谓语是 makes…more independent of…,第二个句子是一个系表结构,is the factor…是表语,在该表语中,which enables a desired value, the setpoint, to be adhered to in the first place 是定语从句,修饰和限定 the factor。因此这句话的意思分成两层:反馈过程使得所谓的控制变量更加独立于外部和内部的干扰变量;反馈过程也是使得期望值即设定值始终保持在初始位置的重要因素。

[3] The "limited" thermostat can as little foresee these influences as it can the occupant's behaviour in opening a window or holding a party with a lot of guests who cause the room to warm up.

本句包含一个比较结构 as little…as…。little 有否定含义,否定的是后面的 foresee(预见)。it can 后省略了动词 foresee。这个结构比较的是不能预见的程度:有限自动温控器不能预知这些影响,就如同不能预知人们开窗和举办聚会的行为。who cause the room to warm up 表面上是 guests 的定语,其实表达的是一种结果:聚会的客人能使屋内温度上升。

[4] Also the behaviour of the system including the effects of disturbance variables which the open-loop control system is able to measure, must be exactly known at all times in order to be able to use the manipulated variable to influence the controlled variable.

本句的主语是 the behaviour of the system,谓语是 must be known,in order to 是目的状语。including the effects of disturbance variables 修饰 system,后面由 which 引导的定语从句修饰

variables。in order to 表示目的,我们可以把它理解为"以便……",也可理解为表示原因和结果的"因此……"。

[5] The behaviour over time of a closed-loop system is normally characterised by the fact that when the input value is abruptly changed, although the output value immediately begins to change, it reaches its end value with some delay.

本句主语为 the behaviour over time of a closed-loop system,谓语是 is characterised by,后续具体"特点"。the fact 里面,大的层次是包含 when 引导的时间状语从句,这其中又插入一个 although 引导的让步状语从句。结构清楚了,理解就不困难了。over time 本意为"随着时间的过去",behaviour over time 意思为"滞后动作"。

[6] In all the different systems discussed which are normally found in combination, we are dealing with so-called linear single variable systems because there is only one output value (= the controlled variable) as well as one input value (= the manipulated variable), and the system possesses linear properties.

本句中,we are dealing with so-called linear single variable system 是主句,because 引导了一个表示原因的状语从句,它由 and 连接的两个并列句组成。in all the different systems 限定 we are dealing with 的范围,同时,which 引导的定语从句表明它是"以组合形式出现"的系统。

Exercises

【Ex. 1】 根据课文内容,回答以下问题。

(1) What is the principle of closed-loop control?

(2) What is the biggest disadvantage of open-loop control?

(3) What is the advantage of open-loop control?

(4) How to classify closed-loop systems?

(5) What is the characteristic of proportional system?

【Ex. 2】 根据下面的英文解释,写出相应的英文词汇。

英 文 解 释	词 汇
a regulating mechanism, as in a vehicle or electric device	
the return of a portion of the output of a process or system to the input, especially when used to maintain performance or to control a system or process	
having no fixed quantitative value	
the act of disturbing	
the degree of hotness or coldness of a body or an environment	
an elastic device, such as a coil of wire, that regains its original shape after being compressed or extended	
a beneficial factor or combination of factors	
a receptacle, such as a carton, can, or jar, in which material is held or carried	
a heating device consisting of a series of connected pipes, typically inside an upright metal structure, through which steam or hot water is circulated so as to radiate heat into the surrounding space	
relative position or rank on a scale	
the act of measuring or the process of being measured	

【Ex. 3】 把下列句子翻译为中文。

(1) Process control is a term applied to the control of variable in a manufacturing process.

(2) Much current work in process control involves extending the use of the digital computer to provide direct digital control(DDC) of the variables.

(3) Numerical control(NC) is a system that uses predetermined instructions called a program to control a sequence of such operations. These instructions are usually stored in the form of numbers —hence the name numerical control.

(4) In direct digital control the computer calculates the values of the manipulated variables directly from the values of the set points and the measurements of the process variables.

(5) Aircraft flight control has been proven to be one of the most complex control applications due to the wide range of system parameters and the interaction between controls.

(6) A linear device is one where output is directly proportional to its input(s) and any dynamic function.

(7) A pneumatic controller, for example, ceases to operate linearly when its output falls to zero or reaches full supply pressure.

(8) By automatic, it generally means that the system is usually capably of adapting to a variety of operating conditions and is able to respond to a class of inputs satisfactorily.

(9) Although open-loop control systems are of limited use, they form the basic elements of an closed loop control systems.

(10) What is missing in the open-loop system for more accurate and more adaptable control is a link or feedback from the output to the input of the system.

【Ex.4】 把下列短文翻译成中文。

Although a servomechanism is not a control application, this device is commonplace in automatic control. A servomechanism, or "servo" for short, is a closed-loop control system in which the controlled variable is mechanical position or motion. It is designed so that the output will quickly and precisely respond to a change in the input command. Thus we may think of a servomechanism as a following device.

【Ex.5】 通过Internet查找资料,借助如"金山词霸"等电子词典和辅助翻译软件,完成以下技术报告。通过E-mail发送给老师,并附上你收集资料的网址。

开环控制和闭环控制的基本知识和典型电路(附上各种典型控制系统的电路框图)。

Text B Digital Control Systems

With the advent of microprocessors and the low cost of computer hardware, greater flexibility and in general better control can be obtained by using a digital computer as the controller.

The structure of a digital computer controlled system is as follows.

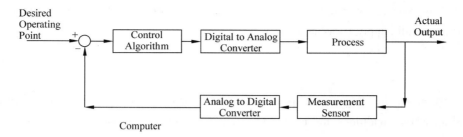

Now we have part of the system as discrete (digital section) and the other as continuous (process and sensors).

Convert digital part to continuous and do mathematics in s or t domain. —difficult, gives transcendental equations to solve.

Convert to discrete and do mathematics in z domain — this is easier but care must be taken. This approach gives no information about signals between samples, and the process being continuous could produce large variations between samples.

In week 1st a simple approximation to obtain the discrete model was presented. It was

$$\frac{dy}{dt} \approx \frac{y(k+1) - y(k)}{\Delta t}$$

and for a 1st order system of the form

$$\frac{dy}{dt} = -ay + u \qquad (1)$$

resulted in the difference equation:

$$y(k+1) = (1 - \alpha\Delta t)y(k) + \Delta t u(k) \qquad (2)$$

Note Δt is the sampling interval, and $u(k)$ is assumed constant during this sampling interval. In the exercises for weeks 1st and 2nd it was shown that even for a constant input, e.g. a unit step, equation (2) y values did not correspond to the y values from equation (1) at the sample points. This was because of the approximation used for the derivative.

Is there a better conversion to discrete form that will be the same at the sampling instances for an input $u(t)$ that is constant during the sampling period? This constant input would be the result of the output from a D/A converter in a digital computer, and hence the following analysis is very relevant to computer controlled systems. Taking L. T. s of equation (1) gives

$$sY(s) - y(0) = -aY(s) + U(s)$$

$$Y(s) = \frac{y(0)}{s+a} + \frac{U(s)}{s+a}$$

Consider the first sample period

where $u(0)$ is constant and hence could be considered as a step of size A. i.e. $U(s) = A/s$

$$Y(s) = \frac{y(0)}{s+a} + \frac{A}{s(s+a)}$$

$$= \frac{y(0)}{s+a} + \frac{k_1}{s} + \frac{k_2}{s+a}$$

$$= \frac{y(0)}{s+a} + \frac{A}{a}\left(\frac{1}{s} - \frac{1}{s+a}\right)$$

Inverse L. T. s gives

$$y(t) = y(0)e^{-at} + \frac{A(1-e^{-at})}{a}, \quad t \geq 0$$

In particular at $t = \Delta t$ or T

$$y(\Delta t) = y(0)e^{-a\Delta t} + \frac{A(1-e^{-a\Delta t})}{a} \tag{3}$$

Noting that $u(t)$ is constant between samples and that the above derivation would be equally valid for any time period, then equation(3) can be generalized to

$$y(k+1) = e^{-a\Delta t}y(k) + \frac{(1-e^{a\Delta t})u(k)}{a} \tag{4}$$

Taking z-transforms gives

$$zY(z) = e^{-a\Delta t}Y(z) + \frac{(1-e^{-a\Delta t})U(z)}{a}$$

giving

$$\frac{Y(z)}{U(z)} = \frac{(1-e^{-a\Delta t})}{a(z-e^{-a\Delta t})} \tag{5}$$

Equation(4) is often written as

$$y(k+1) = e^{-aT}y(k) + \frac{(1-e^{-aT})u(k)}{a} \tag{6}$$

Where T is the sampling period.

1. Example —Car velocity model

Continuous model

$$\frac{v(s)}{m(s)} = \frac{2/15}{s+1/15}$$

For $m(s) = 1/s$

$$v(s) = \frac{2/15}{s(s+1/15)} = \frac{k_1}{s} + \frac{k_2}{s+1/15}$$

Inverse L. T. s gives

$$v(t) = \frac{2}{15}\left[\frac{15}{1}(1-e^{-t/15})\right], \quad t \geq 0$$

$$= 2(1-e^{-t/15}), \quad t \geq 0 \tag{7}$$

Discrete accurate model

$$\frac{v(z)}{m(z)} = \frac{\frac{2}{15} \times \frac{15}{1}(1-e^{-t/15})}{z-e^{-t/15}} = \frac{2(1-e^{-t/15})}{z-e^{-t/15}}$$

If $t = 1\,\text{s}$

$$v(k+1) = 0.936v(k) + 0.128m(k)$$

Discrete approximate model

$$\frac{v(k+1)-v(k)}{1} = -\frac{1}{15}v(k) + \frac{2}{15}m(k)$$

giving

$$v(k+1) = 0.933v(k) + 0.133m(k)$$

Students should simulate the three responses and verify that the discrete accurate conversion response lies exactly on the continuous response and that an error exists between the discrete approximate conversion response and the other two responses.

The above work outlined a more accurate method for finding the discrete model of a continuous system. This approach was only valid for systems with inputs that were constant during the sampling period, and hence is particularly suitable for digital control systems. The method was applied to a first order system, but can readily be extended to higher order systems.

2. PID controller

One of the most successful controllers in industry is known as a PID (Proportional, Integral, Derivative) controller. In the continuous time domain it has the form

$$u(t) = K_p e(t) + K_i \int_0^t e(t)\,dt + K_d \frac{de(t)}{dt} \tag{8}$$

where $e(t)$ is the error signal, and $u(t)$ the control input to the process. In the s-domain

$$U(s) = \left(K_p + \frac{K_i}{s} + K_d s\right)E(s) \tag{9}$$

Note the integral term $K_i \int_0^t e(t)\,dt$ effectively takes into account past errors and uses these to calculate the control $u(t)$ at the present time t. To implement this in a digital computer a discrete approximation is used, and equation (8) becomes

$$u(k) = K_p(e)k + K_i \sum_{j=1}^{k} e(j)\Delta t + K_d \frac{[e(k)-e(k-1)]}{\Delta t} \tag{10}$$

Note the term $\sum_{j=1}^{k} e(j)$ does not have to be calculated for all k since a running sum can be made and updated when the new error is calculated. In the z-domain

$$U(z) = \left(K_p + K_i \frac{T_z}{z-1} + \frac{K_d}{T}\frac{(z-1)}{z}\right)E(z) \tag{11}$$

New Words and Phrases

advent	['ædvənt]	n. (尤指不寻常的人或事)出现,到来
hardware	['hɑːdwɛə]	n. (计算机的)硬件,(电子仪器的)部件
structure	['strʌktʃə]	n. 结构,构造
algorithm	['ælgəriðəm]	n. 运算法则
flexibility	[ˌfleksə'biliti]	n. 弹性,适应性,机动性,挠性

equation	[i'kweiʃən]	n. 方程式,等式
resulted in		导致
domain	[dəu'mein]	n. 范围,领域
transcendental	[trænsen'dentl]	adj. 先验的,超越的,超出人类经验的
convert…to…		把……转换为……
sampling period		采样周期
sampling interval		采样间隔
approach	[ə'prəutʃ]	n. 方法
		vt. 接近,动手处理
		vi. 靠近
sample	['sæmpl]	n. 标本,样品,例子
		vt. 取样,采样,抽取……的样品
approximation	[ə,prɔksi'meiʃən]	n. 近似值,接近,走近
correspond	[kɔris'pɔnd]	vi. 符合,协调,相当,相应
derivative	[di'rivətiv]	n. 导数,微商
D/A		数字/模拟(digital/analog)
converter	[kən'və:tə(r)]	n. 转换器,变换器
relevant	['relivənt]	adj. 有关的,相应的
hence	[hens]	adv. 因此,从此
generalize	['dʒenərəlaiz]	vt. 归纳,概括,推广,普及
simulate	['simjuleit]	vt. 模拟,模仿
verify	['verifai]	vt. 检验,校验,查证,核实
accurate	['ækjurət]	adj. 准确的,精确的
exactly	[ig'zæktli]	adv. 精确地,确切地
valid	['vælid]	adj. 有效的,有根据的,正确的
suitable	['sju:təbl]	adj. 适当的,相配的
transform	[træns'fɔ:m]	vt. 转换,改变,改造,使……变形
		vi. 改变,转化,变换
		n. 变换(式)
velocity	[vi'lɔsiti]	n. 速度,速率,迅速,周转率
outline	['autlain]	n. 大纲,轮廓,略图,外形,要点,概要
		vt. 描画轮廓,略述
proportional	[prə'pɔ:ʃənl]	adj. 比例的,成比例的
PID(Proportional,Integral,Differential(Controller))		比例,积分,微分(控制器)
signal	['signl]	n. 信号

		adj. 信号的
		vt. & vi. 发信号，用信号通知
integral	['intigrəl]	*adj.* 积分的
		n. 积分
implement	['implimənt]	*n.* 工具，器具
		vt. 贯彻，实现
		vt. & vi. 执行

Exercises

【Ex. 6】 根据文章所提供的信息判断正误。

(1) The price of hardware of computer is increasingly high.

(2) A digital computer controlled system consist in control algorithm, digital to analog converter, process, analog digital converter and measurement sensor.

(3) A digital computer controlled system is only designed for discrete variable.

(4) In a digital computer controlled system, digital signal must be converted to analog signal.

(5) According to the equation:
$$y(k+1) = (1 - a\Delta t)y(k) + \Delta t u(k)$$
Δt is the sampling interval.

(6) According to equation(1) and equation(2), if the input does not change, the output will keep constant.

(7) The equation as the following:
$$y(\Delta t) = y(0)e^{-a\Delta t} + \frac{A(1 - e^{-a\Delta t})}{a}$$
only applies in special time period.

(8) According to the passage, the car velocity model is
$$\frac{v(s)}{m(s)} = \frac{2/15}{s + 1/15}$$

(9) PID controller is a very useful controller in industry.

(10) Past errors can be used to calculate the control $u(t)$ at the present time t in a PID controller.

科技英语翻译知识　被动语态的译法

英语中被动语态使用范围很广，主动者经常被省去，科技英语中更是如此。因为科技文献侧重叙事和推理，强调的是作者的观点和发明内容，而不是作者本人。汉语里很少使用被动语态，因此，在翻译时，应尽可能将英语的被动句译成汉语的主动句。视具体情况，也可保留被动句。

1．译成主动句

（1）It is clear that a body can be charged under certain condition.

很显然，在一定条件下物体能够带电。

（原文中的主语 body 在译文中保留为主语。）

（2）Solution to the problem was ultimately found.

这个问题的解决办法终于找到了。

（原文中的 solution 仍然作译文中的主语，不必译成"被找到"。）

（3）If one or more electrons be removed, the atom is said to be positively charged.

如果原子失去一个或多个电子，我们就说该原子带正电荷。

（原文中的主语 electrons 在译文中作宾语。）

（4）Three-phase current should be used for large motors.

大型电动机应当使用三相电流。

（原文中的主语 three-phase current 译作宾语，而原宾语 large motors 在译文中作主语。）

（5）An emf will be generated in the coil.

线圈中会产生电动势。

（地点状语 in the coil 译成主语。）

（6）Any desired voltage may be obtained by means of transformer.

借助变压器可获得任何需要的电压。

（方式状语 by means of transformer 提前，译成无主句。）

（7）The resistance can be tested to determine its voltage and current.

只要知道电压和电流，就能测定电阻。

（原文中的主语 the resistance 在译文中作宾语，译成无主句。）

（8）When these charges move along a wire, that wire is said to carry an electric current.

当这些电荷沿导线运动时，就说该导线带电。

（主语 wire 在汉语句中，与后面的 carry an electric current 构成分句。）

有 it 引导的习惯句型中的被动语态，常有固定的译法：

It is hoped that…希望

It is reported that…据报道

It is said that…据说

It can be foreseen that…可以预料

It cannot be denied that…无可否认

It has been proved that…已经证明

It is arranged that…已经商定，已经准备

It is alleged that…据称

It is announced that…普遍认为

It is believed that…大家相信

It is demonstrated that…据事实

It is accepted that…可接受的是

2. 译成汉语被动句

译成汉语被动句时,除了用"被"字外,还可使用"受"、"由"、"为……所……",等等。

(1) Current will not flow continually, since the circuit is broken by the insulating material.
电流不能继续流动,因为电路被绝缘材料隔断了。

(2) Every charged object is surrounded by an electric field.
每个带电体都为电场所包围。

(3) The magnetic field is produced by an electric current.
磁场由电流产生。

(4) Resistivity is affected by temperature, moisture and structural defects.
电阻率受温度、湿度和结构缺陷的影响。

(5) Current flow is measured in amperes(A) and is represented by the letter I.
电流以安培(A)为单位来量度,并用字母 I 来表示。

(6) Our streets and houses are lighted by electricity.
我们的街道和房屋靠电来照明。

(7) The values of current and voltage required can be obtained from the volt-ampere curve or from an actual circuit.
所需电流和电压的值可以从伏安曲线或实际电路获得。

(8) A graph is often more easily understood than a law.
图比定律更容易为人所理解。

Reading Material

阅读下列文章。

Text	Note

Protel(1)
1. Customizable design environment
  Protel DXP is built on Altium's unique[1] Design Explorer software integration platform[2]. The Design Explorer allows all components of the Protel DXP system to work together and behave as a single design application, providing you with an integrated, customizable[3] user interface that supports the way you want to work.
  The Protel DXP user interface has been extensively redesigned, bringing superior[4] functionality to the design space and a focus on ease-of-use, intuitive control, and consistency of interface across all editors. Through this enhanced interface, Protel DXP brings a powerful new approach to accessing and editing your design data giving you complete control over which objects to view, edit or apply a rule to. Protel DXP also gives you versatile tabular views of all design objects in your project and the ability to easily filter and select multiple objects allows you to make global changes to your design quickly and easily.
  Protel DXP's superb design environment makes design more efficient and lets you choose the most appropriate design path and work the way you want to.
	[1] *adj.* 独特的
	[2] *n.* 平台
	
	[3] *adj.* 用户化的
	
	[4] *adj.* 出众的

Key Features:
- Fully-integrated and intuitive design environment
- Dual monitors support
- Enhanced[5] & consistent interface in every editing environment
- Workspace panels can be docked, floated, or set to automatically hide when not in use
- Floating panels and tool bars will automatically fade out of the way when you are performing any editing in the workspace
- Fully customizable tool bars and views
- Versatile[6] query-driven filtering system to mask, select or zoom the targeted objects
- The new Object Inspector Panel allows you to edit the common properties of all selected objects simultaneously[7]
- A new ListView feature that lets you view your design objects as a tabular list
- The entire work environment can be customized with "click and drag" simplicity[8], and native support for multiple monitors lets you get the best views on your design

You can precisely[9] target design objects for viewing or editing using versatile filtering expressions. Non-targeted objects can be "faded" so you can easily see the parts of the design you want to work with. Multiple[10] objects can be selected and edited simultaneously.

The Protel DXP user interface has been extensively redesigned, bringing superior functionality to the design space and a focus on ease-of-use, intuitive[11] control, and consistency of interface across all editors.

Protel DXP includes a new ListView feature that lets you view your design objects as a tabular[12] list. The ListView can be customized to display various properties of your design objects, and you can sort and filter the data at will to list the objects in any way you choose.

With multiple objects selected, bring up the Object Inspector[13] Panel to edit the common properties of all selected objects simultaneously. The ability to create precisely targeted filtering expressions combined with the power of the Object Inspector makes global editing fast and intuitive.

[5] *adj.* 提高的

[6] *adj.* 通用的

[7] *adv.* 同时地

[8] *n.* 简单,简易

[9] *adv.* 正好

[10] *adj.* 多重的

[11] *adj.* 直觉的

[12] *adj.* 表格式的

[13] *n.* 检测器

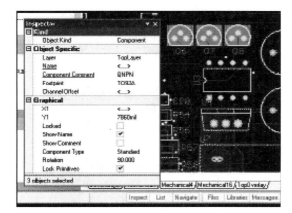

2. Project management and design integrity

Throughout the design process you want to be sure that all your design documents are accurate and reflect the most up-to-date version of the design. Protel DXP not only ensures that your design documents are electrically consistent within themselves, it monitors the integrity of the entire design project.

As well as extensive electrical and design rules checking, Protel DXP maintains the consistency[14] of all source schematics by introducing the concept of project-level "design compilation". The design compiler verifies the electrical and connectivity properties of your source documents. By compiling after changes are made, Protel DXP makes sure that introduced errors are not propagated through to the board design before they are corrected.

Protel DXP also takes the work out of propagating design changes throughout your entire project. A powerful comparator engine highlights design differences between documents and allows you to update either the source schematics or the board design globally, or on a per difference basis, creating a full ECO report of all changes made.

This versatile verification[15] and project-level synchronization system ensures that your entire design project remains error free and in sync throughout the entire design process.

Comparing and managing different versions of your design files is handled easily and efficiently with Protel DXP. Interface directly from Protel's Design Explorer environment to any third-party version control system that supports the popular SCC[16] interface, such as Microsoft Visual SourceSafe®. Protel DXP's extensive interface support for version control allows teams to develop with the confidence that their projects and files will be protected.

Key Features:
- Project-level bi-directional synchronization
- Powerful error checking
- Document comparison features
- Versatile output configurations
- Powerful design verification and debugging[17] at a project-level
- Powerful comparator engine ensures full synchronization between source documents and the target PCB eliminating synchronization problems
- Support for most popular third-party document version control systems using the SCC standard interface
- Uses a Project file to store information about project documents and project-level settings, allowing easy management of all design files and project-level parameters
- Support for multiple board variants[18]

When you compile your design, Protel DXP analyzes your source schematics, checking for a vast array of electrical and drafting errors, and builds a hierarchical[19] electrical and connectivity model of your project — connectivity that can be graphed

[14]	*n.* 一致性
[15]	*n.* 确认
[16]	计算机监控
[17]	*n.* 调试
[18]	*n.* 变量
[19]	*adj.* 分等级的

and verified as soon as the design is compiled.

Protel DXP's powerful comparator engine determines differences across your entire design project and lets you easily reconcile[20] multiple changes in different design documents.

With Protel DXP, design verification extends beyond simple schematic rule checking. Design Verification and Integrity at a project level is a core foundation of Protel DXP's design approach.

3. Design entry

Protel DXP provides a versatile and fully integrated design capture system for both PCB and FPGA applications. There is no limit to the number of sheets, or the depth of design hierarchy. Protel DXP's schematic editor provides an intuitive link between block diagrams and project hierarchy[21]. Each block represents an individual schematic sheet, which can be generated with the click of a button. Wiring these blocks together creates connectivity —connectivity that can be graphed and verified as soon as the design is compiled.

Included with Protel DXP is a comprehensive set of fully integrated component libraries that combine each schematic symbol with their related models — including PCB footprints[22] and Signal Integrity Models. Integrated libraries ensure that model information is always available when needed for any process, such as circuit simulation or a board update. Integrated libraries also offer portability[23], letting you work on your design wherever you have Protel DXP installed. To find the components you want, Protel DXP provides powerful library search features — just open the libraries search dialog from the Libraries panel, and enter the Library type and search criteria[24]. Protel DXP supports previous schematic and PCB library formats, allowing you to use your existing custom Protel libraries within Protel DXP.

Protel DXP fully supports FPGA[25] design flows. Design your FPGA with Protel DXP's powerful schematic editor, using a complete set of macro and primitive libraries supplied for both Xilinx and Altera device families. Generate the EDIF files directly from Protel DXP, ready to load into your FPGA vendor's place and route tools. FPGA projects can be fully integrated into a board design project, with full support for back annotation of FPGA pin out changes to both FPGA and board design projects.

With Protel DXP's true multi-channel design support you only draw the channel schematic once — you then have the full set of features to transfer to PCB layout and step and repeat the routed channel layout on the PCB. Unlike other tools, Protel DXP maintains the channel information in the schematic without the need to flatten the hierarchy, allowing you to update the design or number of channels at any time.

Protel DXP's new project variants system enables you to create a master project and then depopulate it in a spreadsheet[26] — like table to create each assembly variant. You can then generate fabrication, assembly and test information from any variant, or from the master project. This eliminates the need to maintain multiple sets of files for each board variant.

[20] vt. &vi. 使和谐

[21] n. 层次

[22] n. 脚注

[23] n. 可移植

[24] n. 标准

[25] 现场可编程门阵列

[26] n. 电子表格

Key Features: • Design entry for schematic and FPGA applications • Complete set of macro and primitive libraries supplied for both Xilinx and Altera device families • Generate EDIF files directly from schematic • Multi-sheet, hierarchical schematic capture • Unlimited sheet number or hierarchy depth • True multi-channel design support • Support for multiple board variants • Comprehensive set of integrated libraries included: Includes schematic symbols, PCB footprints, Spice models and signal integrity models • Easy navigation[27] design hierarchy and connectivity • Versatile import and export features include direct OrCAD® V9 & V7 schematic and library import • Import/export of AutoCAD ® files up to R14 Protel DXP fully supports FPGA design flows and includes primitive libraries for all current Xilinx and Altera device families. Generate the EDIF files directly from Protel DXP, ready to load into your FPGA vendor's4[28] place and route tools. With Protel DXP's true multi-channel design support you only draw the channel schematic once —you then have the full set of features to transfer to PCB layout and step and repeat the routed channel layout on the PCB. Included with Protel DXP is a comprehensive[29] set of fully integrated component libraries that combine each schematic symbol with their related models — including PCB footprints and Signal Integrity Models. Protel DXP's schematic editor is easy to use, yet capable of managing even the largest design. There is no limit to the number of sheets, or the depth of design hierarchy.	[27] n. 导航,向导 [28] n. 供应商 [29] adj. 全面的,广泛的

参考译文 控制基础

1. 闭环控制原理

闭环控制的原理就是通过控制器测量反馈回来的控制值,并与设定值一起输入控制系统。反馈过程使得所谓的控制变量更加独立于外部和内部的干扰变量,它也是使得期望值即设定值始终被保持在起始位置的重要因素。当由控制器输出的操作变量影响控制变量时,所谓的控制环会适时地关闭。

技术系统可以处理多种控制变量,例如:电流、电压、温度、压力、流量、旋转速度、转动角、化学浓度和许多别的对象。干扰变量也有物理属性。

这些控制环的术语可以用大家熟悉的利用冷却控制器进行室温控制的例子来解释:为

了将室温保持在22℃。温度可以通过可旋转的旋钮进行设定（设定值）。温度（控制变量）是由传感器测量的。室温同设定值之间的偏差由内置的控制器测得——经常使用双金属簧片（控制偏差），然后再用于开阀和关阀（操作变量）。

什么是干扰变量？首先是室外温度的影响和通过窗户的阳光的照射。"有限"自动温控器不能预知这些影响，就如同不能预知人们开窗和举办聚会的行为，聚会的客人会使屋内温度上升一样。总之，控制器可以补偿由一种或多种干扰变量造成的影响，尽管有些延时，还是能使温度重新回到期望的水平。

2. 开环控制原理

没有建立闭环控制的地方就有开环控制。和闭环控制相比，开环控制的最大不足就是，对于不知道或不能测量的干扰变量无法进行补偿。也就是系统的动作，包括开环系统能够测量的干扰变量的影响，始终要精确了解以便使用操作变量影响控制变量。

开环控制系统的一个优点是，由于没有反馈而使系统稳定——这正是闭环系统的问题所在。

3. 闭环控制系统的分类

闭环控制系统不是按照被控对象的物理值，而是根据它的滞后动作进行分类的。容器内的水平面能够像电容的电压一样精确地进行数学描述。

动作滞后能够通过，比如说，突然改变输入值，然后观察输出值的变化确定。根据基本的物理定律的知识可以充分估计这种动作。只有较少情况才需要计算。

闭环控制系统的滞后动作的特点通常为：当输入值突然变化时，尽管输出值立即就开始变化，但达到最终的值需要一些延时。

闭环系统可以根据是否有自调节进一步区分。

在有自调节的系统中，当输入值突然变化后，经过一段时间后输出又表现为一个恒定值。这种系统常称为比例系统或 P 系统。举一个加热环的例子：输入值是加热电能，输出值是环的温度。

在没有自调节的系统中，当输入值突然变化后输出值也会上升或下降。只有当输入在零点时输出才保持恒定。

这种系统一般称为积分系统或 I 系统。容器的液面控制就是这样的例子：注入的流量是输入值，液体的水平高度是输出值。

另一种重要的系统类型是有死区的系统。这种情况下输入的改变只有经过死区时间延迟后才能在输出上反应出来。在技术系统中，设定位置与测量位置之间的距离产生了死区时间。例如传送带：传送带开始段传送物的量等于输入值，传送带末端的量等于输出值。死区的时间可以通过传送带的长度除以传送速度得到，因此它可以是变化的。

这里讨论的不同的系统通常以组合形式出现。我们讨论的是所谓的线性单变量系统，因为只有一个输出值（控制量），也只有一个出入值（操纵量），并且系统具有线性属性。对这些非常普通的系统的控制就被称为单变量控制系统。

Unit 8

Text A PLC

Programmable Logic Controllers (PLCs), also referred to as programmable controllers, are in the computer family. They are used in commercial and industrial applications. A PLC monitors inputs, makes decisions based on its program, and controls outputs to automate a process or machine.[1] This course is meant to supply you with basic information about the functions and configurations of PLCs.

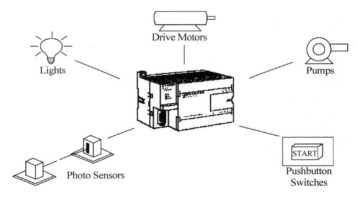

1. Basic PLC Operation

PLCs consist of input modules or points, a Central Processing Unit (CPU), and output modules or points. An input accepts a variety of digital or analog signals from various field devices (sensors) and converts them into logic signals that can be used by the CPU.[2] The CPU makes decisions and executes control instructions based on program instructions in memory. Output modules convert control instructions from the CPU into a digital or analog signal that can be used to control various field devices (actuators). A programming device is used to input the desired instructions. These instructions determine what the PLC will do for a specific input. An operator interface allows process information to be displayed and new control parameters to be entered. Pushbuttons (sensors), in this simple example, connected to PLC inputs, can be used to start and stop a motor connected to a PLC through a motor starter (actuator).

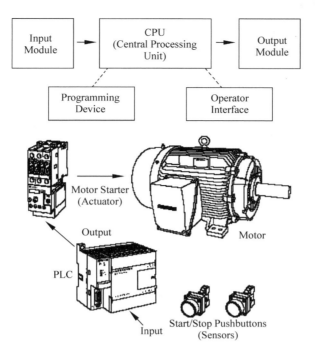

2. Hard-Wired Control

Prior to PLCs, many of these control tasks were solved by contactors or relay controls. This is often referred to as hardwired control. Circuit diagrams had to be designed, electrical components specified and installed, and wiring lists created. [3] Electricians would then wire the components necessary to perform a specific task. If an error was made the wires had to be reconnected correctly. A change in function or system expansion required extensive component changes and rewiring.

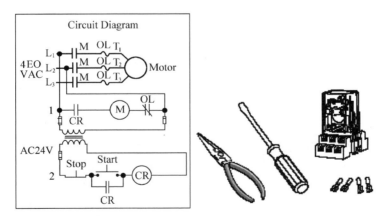

3. Advantages of PLCs

The same, as well as more complex tasks, can be done with a PLC. Wiring between devices and relay contacts is done in the PLC program. Hard-wiring, though still required to connect field devices, is less intensive. Modifying the application and correcting errors are easier to handle. It is easier to create and change a program in a PLC than it is to wire and rewire a circuit.

Following are just a few of the advantages of PLCs:
- Smaller physical size than hard-wire solutions.
- Easier and faster to make changes.
- PLCs have integrated diagnostics and override functions.
- Applications can be immediately documented.
- Applications can be duplicated faster and less expensively.

4. Siemens PLCs

Siemens makes several PLC product lines in the SIMATIC® S7 family. They are: S7-200, S7-300, and S7-400.

(1) S7-200

The S7-200 is referred to as a micro-PLC because of its small size. The S7-200 has a brick design which means that the power supply and I/O terminal are on-board. The S7-200 can be used on smaller, stand-alone applications such as elevators, car washes, or mixing machines. It can also be used on more complex industrial applications such as bottling and packaging machines.

(2) S7-300 and S7-400

The S7-300 and S7-400 PLCs are used in more complex applications that support a greater number of I/O points. Both PLCs are modular and expandable. The power supply and I/O consist of separate modules connected to the CPU. Choosing either the S7-300 or S7-400 depends on the complexity of the task and possibility of future expansion. Your Siemens sales representative can provide you with additional information on any of the Siemens's PLCs.

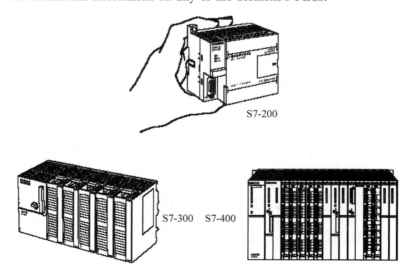

5. CPU

CPU is a microprocessor system that contains the system memory and is the PLC decision-making unit. The CPU monitors the inputs and makes decisions based on instructions held in the program memory. The CPU performs relay, counting, timing, data comparison, and sequential operations.

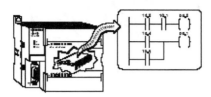

6. Programming Devices

The program is created in a programming device (PG) and then transferred to the PLC. The program for the S7-200 can be created using a dedicated Siemens SIMATIC S7 programming device, such as PG 720 or PG 740, if STEP 7 Micro/WIN software is installed.

A personal computer (PC), with STEP 7 Micro/WIN installed, can also be used as a programming device for the S7-200.

7. Software

A software program is required in order to tell the PLC what instructions it must follow. Programming software is typically PLC specific. A software package for one PLC, or one family of PLCs, such as the S7 family, would not be used on other PLCs. The S7-200 uses a Windows-based software program called STEP 7-Micro/WIN32. The PG 720 and PG 740 have STEP 7-Micro/WIN32 software pre-installed. Micro/WIN32 is installed on a personal computer in a similar manner to any other computer software.

8. Connector Cables PPI (Point-to-Point Interface)

Connector cables are required to transfer data from the programming device to the PLC. Communication can only take place when two devices speak the same language or protocol. Communication between a Siemens programming device and a S7-200 is referred to as PPI protocol. An appropriate cable is required for a programming device such as PG 720 or PG 740. The S7-200 uses a 9-pin, D-connector. This is a straight-through serial device that is compatible with Siemens programming devices (MPI port) and is a standard connector for other serial interfaces.

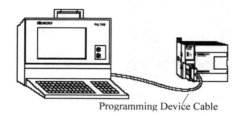

Programming Device Cable

A special cable, referred to as a PC/PPI cable, is needed when a personal computer is used as a programming device. This cable allows the serial interface of the PLC to communicate with the RS-232 serial interface of a personal computer. DIP switches on the PC/PPI cable are used to select an appropriate speed (bit rate) at which information is passed between the PLC and the computer.[4]

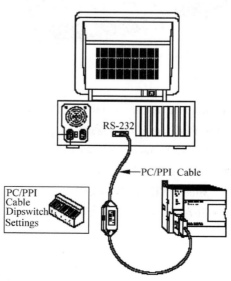

New Words and Phrases

PLC (Programmable Logic Controller)		可编程序逻辑控制器
programmable	['prəugræməbl]	*adj.* 可设计的, 可编程的
monitor	['mɔnitə]	*n.* 监听器, 监视器, 监控器 *vt. & vi.* 监控
configuration	[kən,figju'reiʃən]	*n.* 构造, 结构, 配置, 外形
field	[fi:ld]	*n.* 原野, 现场
module	['mɔdju:l]	*n.* 模块
digital signal		数字信号
analog signal		模拟信号
convert	[kən'və:t]	*n.* 转换 *vt.* 使转变, 转换……
starter	['sta:tə]	*n.* 启动器, 启动钮
actuator	['æktjueitə]	*n.* 操作[执行]机构, 执行器
Hard-Wired Control		硬接线控制
reconnect	[,ri:kə'nekt]	*vt. & vi.* 再接合
complex	['kɔmpleks]	*adj.* 复杂的, 合成的, 综合的
handle	['hændl]	*vt.* 触摸, 运用, 买卖, 处理, 操作 *vi.* 搬运, 易于操纵
interface	['intə(:),feis]	*n.* 分界面, <计>界面, 接口
relay	['ri:lei]	*n.* 继电器
extensive	[iks'tensiv]	*adj.* 广大的, 广阔的, 广泛的
diagnostics	[,daiəg'nɔstiks]	*n.* 诊断学
override	[,əuvə'raid]	*vt.* 不顾, 无视, 蔑视
Siemens	['si:mənz]	西门子(德国电气公司)
SIMATIC		产品名
elevator	['eliveitə]	*n.* 电梯, 升降机, <空>升降舵
bottling	['bɔtliŋ]	*n.* 装瓶(子)
packaging	['pækidʒiŋ]	*n.* 包装
microprocessor	[maikrəu'prəusesə]	*n.* 微处理器
PPI (Point-to-Point Interface)		点对点接口
compatible	[kəm'pætəbl]	*adj.* 谐调的, 一致的, 兼容的
brick	[brik]	*n.* 砖, 砖块, 砖形物(如茶砖, 冰砖等)
I/O point		输入/输出点
depend on		依靠, 依赖

counting	['kauntiŋ]		计数
timing	['taimiŋ]	n.	计时,定时
data comparison			数据比较
PG			编程器(编程设备)
STEP 7			一种 PLC 编程语言
protocol	['prəutəkɔl]	n.	协议
bit rate		n.	比特率

Notes

[1] A PLC monitors inputs, makes decisions based on its program, and controls outputs to automate a process or machine.

本句是个简单句。主语是 PLC,谓语由 monitors,makes,controls 组成,分别描写三个行为,不定式短语 to automate a process or machine 作为目的状语。based on its program 修饰 makes decisions,告诉我们结果是"根据程序"得到的。

[2] An input accepts a variety of digital or analog signals from various field devices (sensors) and converts them into logic signals that can be used by the CPU.

本句结构很简单,由 and 连接两个并列句,后一个并列句里 that 引导的定语从句修饰 signal。an input 里的 an 是不定冠词,泛指 input,而不是指"一个"input。field devices 作现场设备解释。

[3] Circuit diagrams had to be designed, electrical components specified and installed, and wiring lists created.

本句是英语常见的动词省略现象。本句的前半句谓语是完整的:had to be designed。后面两个部分,分别以 electrical components 以及 wiring lists 作主语,谓语部分的 had to be 被省略。这句话补充完整应该是 Circuit diagrams had to be designed, electrical components had to be specified and installed, and wiring lists had to be created.

[4] DIP switches on the PC/PPI cable are used to select an appropriate speed (bit rate) at which information is passed between the PLC and the computer.

本句中有一个以 which 引导的定语从句。在这个定语从句中,介词 at 前置,和引导词在一起,实际指的是 at the speed,这个 speed 指的是 PLC 和计算机之间信息传输的速度。

Exercises

【Ex.1】 根据课文内容,回答以下问题。

(1) How does a PLC work?

(2) What does a PLC consist of?

(3) What are the advantages of PLCs?

(4) For Siemens PLCs, can a personal computer be used as a programming device(PG)?

(5) When a personal computer is used as a programming device, what cable is needed to connect PLC?

【Ex.2】 根据下面的英文解释,写出相应的英文词汇。

英 文 解 释	词 汇
programmable Logic Controller	
a usual electronic device used to record, regulate, or control a process or system	
a standardized, often interchangeable component of a system or construction that is designed for easy assembly or flexible use	
a device, such as a photoelectric cell, that receives and responds to a signal or stimulus	
a surface forming a common boundary between adjacent regions, bodies, substances, or phases	
a device that responds to a small current or voltage change by activating switches or other devices in an electric circuit	
a platform or an enclosure raised and lowered in a vertical shaft to transport people or freight	
an integrated circuit that contains the entire central processing unit of a computer on a single chip	
a usual pliable metallic strand or rod made in many lengths and diameters, sometimes clad and often electrically insulated, used chiefly for structural support or to conduct electricity	
a molded rectangular block of clay baked by the sun or in a kiln until hard and used as a building and paving material	

【Ex.3】 把下列句子翻译为中文。

(1) A super capacitor, so named because of its ability to maintain a charge for a long period of time, protects data stored in RAM in the event of a power loss.

(2) Output devices, such as relays, are connected to the terminal strip under the top cover of the PLC.

(3) High-speed instructions allow for events and interrupts to occur independent of the PLC scan time.

（4）One of the most confusing aspects of PLC programming for first-time users is the relationship between the device that controls a status bit and the programming function that uses a status bit.

（5）Once a program has been written it needs to be tested and debugged.

（6）A limit switch could be used to stop the motor or prevent the motor from being started.

（7）In addition, analog modules are available for use with thermocouple and RTD type sensors used in to achieve a high level of accuracy in temperature measurement.

（8）Analog outputs are used in applications requiring control capability of field devices which respond to continuous voltage or current levels.

（9）The transducer takes the voltage signal and, depending on the requirement, amplifies, reduces, or changes it into another signal which controls the device.

（10）PROFIBUS DP is an open bus standard for a wide range of applications in various manufacturing and automation processes.

【Ex. 4】 把下列短文翻译成中文。

The programming software can be run Off-line or On-line. Off-line programming allows the user to edit the ladder diagram and perform a number of maintenance tasks. The PLC does not need to be connected to the programming device in this mode. On-line programming requires the PLC to be connected to the programming device. In this mode program changes are downloaded to the PLC. In addition, status of the I/O elements can be monitored. The CPU can be started, stopped, or rested.

【Ex.5】 通过 Internet 查找资料,借助如"金山词霸"等电子词典和辅助翻译软件,完成以下技术报告。通过 E-mail 发送给老师,并附上你收集资料的网址。

(1) 当前世界上有哪些最主要的 PLC 生产厂家以及有哪些最新型的产品(附上各种最新产品的图片)。

(2) 当前有哪些最主要的 PLC 编程语言(附上各种编程语言的操作界面)。

Text B SPICE

SPICE (Simulation Program with Integrated Circuit Emphasis) is a general-purpose, open source analog electronic circuit simulator. It is a program used in integrated circuit and board-level design to check the integrity of circuit designs and to predict circuit behavior.

1. Introduction

Unlike board-level designs composed of discrete parts, it is not practical to breadboard integrated circuits before manufacture. Further, the high costs of photolithographic masks and other manufacturing prerequisites make it essential to design the circuit to be as close to perfect as possible before the integrated circuit is first built. Simulating the circuit with SPICE is the industry-standard way to verify circuit operation at the transistor level before committing to manufacturing an integrated circuit.

Board-level circuit designs can often be breadboarded for testing. Even with a breadboard, some circuit properties may not be accurate compared to the final printed wiring board, such as parasitic resistances and capacitances. These parasitic components can often be estimated more accurately using SPICE simulation. Also, designers may want more information about the circuit than is available from a single mock-up. For instance, circuit performance is affected by component manufacturing tolerances. In these cases it is common to use SPICE to perform Monte Carlo simulations of the effect of component variations on performance, a task which is impractical using calculations by hand for a circuit of any appreciable complexity.

Circuit simulation programs, of which SPICE and derivatives are the most prominent, take a text netlist describing the circuit elements (transistors, resistors, capacitors, etc.) and their connections, and translate this description into equations to be solved. The general equations produced are nonlinear differential algebraic equations which are solved using implicit integration methods, Newton's method and sparse matrix techniques.

2. Transient analysis

Since transient analysis is dependent on time, it uses different analysis algorithms, control options with different convergence-related issues and different initialization parameters than DC analysis. However, since a transient analysis first performs a DC operating point analysis (unless the UIC option is specified in the .TRAN statement), most of the DC analysis algorithms, control options, and initialization and convergence issues apply to transient analysis.

Some circuits, such as oscillators or circuits with feedback, do not have stable operating point solutions. For these circuits, either the feedback loop must be broken so that a DC operating point can be calculated or the initial conditions must be provided in the simulation input. The DC operating point analysis is bypassed if the UIC parameter is included in the .TRAN statement. If UIC is included in the .TRAN statement, a transient analysis is started using node voltages specified in an .IC statement. If a node is set to 5 V in a .IC statement, the value at that node for the first time point (time 0) is 5 V.

You can use the .OP statement to store an estimate of the DC operating point during a transient analysis.

.TRAN 1ns 100ns UIC .OP 20ns

The .TRAN statement UIC parameter in the above example bypasses the initial DC operating point analysis. The .OP statement calculates transient operating point at t = 20 ns during the transient analysis.

Although a transient analysis might provide a convergent DC solution, the transient analysis itself can still fail to converge. In a transient analysis, the error message "internal timestep too small" indicates that the circuit failed to converge. The convergence failure might be due to stated initial conditions that are not close enough to the actual DC operating point values.

3. Program features and structure

SPICE became popular because it contains the analyses and models needed to design integrated circuits of the time, and was robust enough and fast enough to be practical to use. Precursors to SPICE often had a single purpose: The BIAS program, for example, did simulation of bipolar transistor circuit operating points; the SLIC (simulator for linear integrated circuits) program did only small-signal analyses. SPICE combined operating point solutions, transient analysis, and various small-signal analyses with the circuit elements and device models needed to successfully simulate many circuits.

3.1 Analyses

SPICE2 included these analyses:
- AC analysis (linear small-signal frequency domain analysis)
- DC analysis (nonlinear quiescent point calculation)
- DC transfer curve analysis (a sequence of nonlinear operating points calculated while sweeping an input voltage or current, or a circuit parameter)
- Noise analysis (a small signal analysis done using an adjoint matrix technique which sums uncorrelated noise currents at a chosen output point)
- Transfer function analysis (a small-signal input/output gain and impedance calculation)
- Transient analysis (time-domain large-signal solution of nonlinear differential algebraic equations)

Since SPICE is generally used to model nonlinear circuits, the small signal analyses are necessarily preceded by a quiescent point calculation at which the circuit is linearized. SPICE2 also contained code for other small-signal analyses: sensitivity analysis, pole-zero analysis, and small-signal distortion analysis. Analysis at various temperatures was done by automatically updating semiconductor model parameters for temperature, allowing the circuit to be simulated at temperature extremes.

Other circuit simulators have added many analyses beyond those in SPICE2 to address changing industry requirements. Parametric sweeps were added to analyze circuit performance with changing manufacturing tolerances or operating conditions. Loop gain and stability calculations were added for analog circuits. Harmonic balance or time-domain steady state analyses were added for RF and switched-capacitor circuit design. However, a public-domain circuit simulator containing the modern analyses and features needed to become a successor in popularity to SPICE has not yet emerged.

It is very important to use appropriate analyses with carefully chosen parameters. For example, application of linear analysis to nonlinear circuits should be justified separately. Also, application of transient analysis with default simulation parameters can lead to qualitatively wrong conclusions on circuit dynamics.

3.2 Device models

SPICE2 included many semiconductor device compact models: three levels of MOSFET model, a combined Ebers – Moll and Gummel-Poon bipolar model, a JFET model, and a model for a junction diode. In addition, it had many other elements: resistors, capacitors, inductors (including coupling), independent voltage and current sources, ideal transmission lines, active components and voltage and current controlled sources.

SPICE3 added more sophisticated MOSFET models, which were required due to advances in semiconductor technology. In particular, the BSIM family of models were added, which were also developed at UC Berkeley.

Commercial and industrial SPICE simulators have added many other device models as technology advanced and earlier models became inadequate. To attempt standardization of these models so that a set of model parameters may be used in different simulators, an industry working group was formed, the Compact Model Council, to choose, maintain and promote the use of standard models. The standard models today include BSIM3, BSIM4, BSIMSOI, PSP, HICUM, and MEXTRAM.

3.3 Input and output: Netlists, schematic capture and plotting

SPICE2 took a text netlist as input and produced line printer listings as output, which fit with the computing environment in 1975. These listings were either columns of numbers corresponding to calculated outputs (typically voltages or currents), or line-printer character "plots". SPICE3 retained the netlist for circuit description, but allowed analyses to be controlled from a command line interface similar to the C shell. SPICE3 also added basic X plotting, as UNIX and engineering workstations became common.

Vendors and various free software projects have added schematic capture front-ends to SPICE, allowing a schematic diagram of the circuit to be drawn and the netlist to be automatically generated. Also, graphical user interfaces were added for selecting the simulations to be done and manipulating the voltage and current output vectors. In addition, very capable graphing utilities have been added to see waveforms and graphs of parametric dependencies. Several free versions of these extended programs are available, some as introductory limited packages, and some without restrictions.

New Words and Phrases

SPICE (Simulation Program with Integrated Circuit Emphasis)		集成电路模拟程序
open source		开放源码,开源
analog electronic circuit		模拟电子电路
simulator	['simjuleitə]	n. 模拟器
integrated circuit		集成电路
board-level design		板级设计
integrity	[in'tegriti]	n. 完整性
behavior	[bi'heivjə]	n. 行为
discrete	[dis'kri:t]	adj. 不连续的,分散的,离散的
breadboard	['bredbɔ:d]	n. 电路实验板
photolithographic	[,fəutə,liθə'græfik]	adj. 照相平版印刷(法)的
photolithographic masks		光刻掩膜
printed wiring board		印刷线路板
parasitic	[,pærə'sitik]	adj. 寄生的
mock-up	['mɔkʌp]	n. 实验或教学用的实物大模型
manufacturing tolerance		制造公差
Monte Carlo	[mɔnti 'kɑ:ləu]	n. 蒙特卡洛
impractical	[im'præktikəl]	adj. 不切实际的,昧于实际的
complexity	[kəm'pleksiti]	n. 复杂性
derivative	[di'rivətiv]	adj. 引出的
		n. 派生的事物
netlist	[netlist]	n. 连线表,网络表
circuit element		电路元件
differential algebraic equation		微分代数方程
implicit integration methods		隐式积分方法
Newton's method		牛顿法
sparse matrix		稀疏矩阵
transient analysis		瞬态分析,暂态分析
convergence	[kən'və:dʒəns]	n. 收敛
initialization parameter		最初参数,初始参数
UIC (Use Initial Conditions)		使用初始条件
initialization	[i,niʃəlai'zeiʃən]	n. 设定初值,初始化

feedback loop			反馈回路,反馈环
statement	['steitmənt]	n.	语句
bypass	['baipɑ:s]	n.	旁路
		vt.	绕过,设旁路,迂回
convergent	[kən'və:dʒnet]	adj.	会集于一点的,会聚性的,收敛的
of the time			当时的,当代的
robust	[rə'bʌst]	adj.	健壮的
precursor	[pri(:)'kə:sə]	n.	先驱
bipolar transistor			双极(性)晶体管,场效应晶体管
SLIC (Simulator for Linear Integrated Circuits)			线性集成电路模拟程序
linear small-signal frequency domain analysis			线性小信号频域分析
nonlinear quiescent point calculation			非线性静态点计算
transfer curve			转换曲线
sweeping	['swi:piŋ]	n.	扫描
noise analysis			噪声分析
adjoint matrix			伴随矩阵
transfer function			传递函数,转移函数
time-domain large-signal solution			时域大信号解
linearize	['liniə,raiz]	vt.	使线性化
sensitivity analysis			灵敏度分析
pole-zero analysis			极零点分析
distortion	[dis'tɔ:ʃən]	n.	变形,失真
temperature extremes			温度极限
parametric sweep			参数扫描
loop gain			环路增益
stability calculation			稳定计算,稳性计算
harmonic balance			谐波平衡
steady state			恒稳态,定态
successor	[sək'sesə]	n.	继承者,接任者
default	[di'fɔ:lt]	n.	默认(值)
compact model			精简模型,简化模型
MOSFET (Metallic Oxide Semiconductor Field Effect Transistor)			金属氧化物半导体场效应晶体管
JFER (Junction Field Effect Transistor)			结栅场效应晶体管
junction diode			面结型二极管
coupling	['kʌpliŋ]	n.	接合,耦合
transmission lines			传输线,波导线

semiconductor technology		半导体工艺
BSIM (Berkeley Short – channel IGFET Model)		伯克利短沟道 IGFET 模型
inadequate	[in'ædikwit]	adj. 不充分的，不适当的
standardization	[,stændədai'zeiʃən]	n. 标准化
schematic capture		获得原理图
plot	[plɔt]	vt. 绘图
line printer		行式打印机
character	['kæriktə]	n. 字符
command line		命令行
workstation	['wəːksteiʃ(ə)n]	n. 工作站
schematic diagram		原理图，示意图
graphical user interface		图形用户界面
vector	['vektə]	n. 向量，矢量
graph	[grɑːf]	n. 图表，曲线图

Exercises

【Ex. 6】 根据文章所提供的信息判断正误。

(1) SPICE is a specific-purpose, open source analog electronic circuit simulator.

(2) Board-level circuit designs can often be breadboarded for testing.

(3) Board-level circuit designers need less information about the circuit than is available from a single mock-up.

(4) You can use the .TRAN statement to store an estimate of the DC operating point during a transient analysis.

(5) SPICE contained the analyses and models needed to design integrated circuits of the time.

(6) SPICE is generally used to model linear circuits.

(7) Loop gain and stability calculations were added for analog circuits.

(8) Qualitatively wrong conclusions on circuit dynamics can result from application of transient analysis with default simulation parameters can lead to.

(9) The standard models today include BSIM3, BSIM4, BSIMSOI, PSP, HICUM, and MEXTRAM.

(10) SPICE2 produced line printer listings as output, and these listings were only columns of numbers corresponding to calculated outputs (typically voltages or currents).

科技英语翻译知识　从句的译法

英语的从句分为定语从句、主语从句、宾语从句、状语从句、表语从句以及同位语从句。由于英汉两种语言结构的不同，翻译时，应该根据不同结构、不同含义采用不同译法。下面一一讨论。

1．定语从句翻译法

1）合译法

把定语从句放在被修饰的词语之前，从而将英语复合句翻译成汉语单句。如：

（1）All substances which can conduct electricity are called conductors.

一切能导电的物质叫做导体。

（不能说"一切物质叫做导体，该物质能导电"。第一分句在意义上不能独立存在，所以要合起来翻译。）

（2）An electric motor changes electrical energy into mechanical energy that can be used to do work.

电动机能把电能变成可用来做功的机械能。

（此句的定语从句也可分开来译，但它较短，还是合译为好。）

（3）New electron tubes could be built that worked at much higher voltages.

如今制造的电子管，工作电压要高得多。

（这一句定语从句是全句的重点，将从句顺序译成简单句中的谓语，从而突出从句的内容。）

2）分译法

根据定语从句的不同情况，可以将其翻译成并列分句、其他从句或词组等。如：

（1）Mechanical energy is changed into electric energy, which in turn is changed into mechanical energy.

机械能转变为电能，而电能又转变为机械能。

（本句的翻译被拆成两个分句，变成转折分句。）

（2）This, of course, includes the movement of electrons, which are negatively charged particles.

当然，这包括电子（带负电的粒子）的运动。

（定语从句在本句的翻译中转变为词组。）

（3）The strike would prevent the docking of ocean steamships which require assistance of tugboats.

罢工会使远洋航船不能靠岸，因为他们需要拖船的帮助。

（翻译成原因状语从句。）

（4）A geological prospecting engineer who had made a spectral analysis of ores discovered a new open-cut coalmine.

一位地质勘探工程师对光谱进行了分析之后，发现了新的露天煤矿。

（翻译为时间状语结构。）

（5）The delivery of public services has tended to be an area where we decorate an obsolete process with technology.

公共服务的提供方式已趋陈旧，这正是我们必须采用技术加以装备的领域。

（翻译为并列分句。）

（6）We now live in a very new economy, a service economy, where relationships are becoming more important than physical products.

现在我们正生活于一种全新的经济,即服务性经济中,各种关系越来越比物质产品更为重要。

(翻译为并列分句。)

2. 主语从句翻译法

1)"的"字结构

以 that,what,who,where,whatever 等代词引导的主语从句,可以将从句翻译成"的"字结构,如:

(1) It is important that science and technology be pushed forward as quickly as possible.

重要的是要把科学技术搞上去。

(2) Whoever breaks the law will be punished.

凡是犯法的人都要受到法律的制裁。

(主语从句与主句合译成简单句,按顺序译出。)

2)"主-谓-宾"结构

以上代词引导的主语从句也可以译成"主-谓-宾"结构,从句本身作句子的主语,其余部分按原文顺序译出,如:

Whether the Government should increase the financing of pure science at the expense of technology or vice versa often depends on the issue of which is seen as the driving force.

政府究竟是以牺牲对技术的经费投入来增加对纯理论科学的经费投入,还是相反,这往往取决于把哪一方看作是驱动的力量。

3)分译法

把原来的状语从句从整体结构中分离出来,译成另一个相对独立的单句,如:

It has been rightly stated that this situation is a threat to international security.

这个局势对国际安全是个威胁,这样的说法是完全正确的。

(It 是形式主语,that this situation is a threat to international security 是真正的宾语。)

3. 宾语从句翻译法

由 that,what,how,where 等词引导的宾语从句一般按照原文顺序翻译,即顺译法,如:

(1) Scientists have reason to think that a man can put up with far more radiation than 0.1 rem without being damaged.

科学家们有理由认为人可以忍受远超过 0.1 rem 的辐射而不受伤害。

(2) We wish to inform you that we specialize in the export of Chinese textiles and shall be glad to enter into business relations with you on the basis of equality and mutual benefit.

我公司专门办理中国纺织品出口业务,并愿在平等互利的基础上同贵公司建立业务关系。

4. 状语从句翻译法

有时也可以译为并列句,如:

Electricity is such an important energy that modern industry could not develop without it.

电是一种非常重要的能量,没有它,现代化工业就不能发展。

(原文由 such...that...引导的结果状语从句译为汉语的并列句。)

5. 表语从句翻译法

大部分情况下可以采用顺译法,也可以用逆译法,如:

(1) My point is that the frequent complaint of one generation about the one immediately following it is inevitable.

我的观点是一代人经常抱怨下一代人是不可避免的。

(顺译法)

(2) His view of the press was that the reporters were either for him or against him.

他对新闻界的看法是,记者们不是支持他,就是反对他。

(顺译法)

(3) Water and food is what the people in the area are badly needed.

该地区的人们最需要的是水和食品。

(逆译法)

6. 同位语从句翻译法

先翻译从句,即从句前置,如:

(1) This is a universally accepted principle of international law that the territory sovereignty does not admit of infringement.

一个国家的领土不容侵犯,这是国际法中尽人皆知的准则。

(2) Despite the fact that comets are probably the most numerous astronomical bodies in the solar system aside from small meteor fragments and the asteroids, they are largely a mystery.

在太阳系中除小片流星和小行星外,彗星大概是数量最多的天体了,尽管如此,它们仍旧基本上是神秘莫测的。

Reading Material

阅读下列文章。

Text	Note
PSpice Tutorial 1. Opening PSpice • Find PSpice on the C-Drive. Open Schematics or you can go to PSpice A_D and then click on the schematic icon[1] 2. Drawing the circuit (1) Getting the Parts • The first thing that you have to do is get some or all of the parts you need. • This can be done by	[1] *n.* 图标

- Clicking on the "get new parts" button ▦ , or
   - Pressing "Control + G", or
   - Going to "Draw" and selecting "Get New Part…"
 • Once this box is open, select a part that you want in your circuit. This can be done by typing in the name or scrolling[2] down the list until you find it.
 • Some common parts are
  - r - resistor
  - C - capacitor
  - L - inductor[3]
  - d - diode[4]
  - GND_ANALOG or GND_EARTH —— this is very important, you MUST have a ground in your circuit
  - VAC and VDC
 • Upon selecting your parts, click on the place button then click where you want it placed (somewhere on the white page with the blue dots). Don't worry about putting it in exactly the right place, it can always be moved later.
 • Once you have all the parts you think you need, close that box. You can always open it again later if you need more or different parts.
(2) Placing the Parts
 • You should have most of the parts that you need at this point.
 • Now, all you do is put them in the places that make the most sense[5] (usually a rectangle works well for simple circuits). Just select the part and drag it where you want it.
 • To rotate parts so that they will fit in your circuit nicely[6], click on the part and press "Ctrl + R" (or Edit "Rotate"). To flip them, press "Ctrl + F" (or Edit "Flip").
 • If you have any parts left over, just select them and press "Delete".
(3) Connecting the Circuit
 • Now that your parts are arranged well, you'll have to attach them with wires.
 • Go up to the tool bar and
  - select "Draw Wire" ▦ or
  - "Ctrl + W" or
  - go to "Draw" and select "Wire".
 • With the pencil looking pointer, click on one end of a part, when you move your mouse around, you should see dotted lines appear. Attach the other end of your wire to the next part in the circuit.
 • Repeat this until your circuit is completely wired.
 • If you want to make a node (to make a wire go more than one place), click somewhere on the wire and then click to the part (or the other wire). Or you can go from the part to the wire.
 • To get rid of [7] the pencil, right click.
 • If you end up with extra dots near your parts, you probably have an extra wire, select this short wire (it will turn red), then press "Delete".
	[2] *vt. & vi.* 滚动
	
	[3] *n.* 电感
	[4] *n.* 二极管
	
	[5] *n.* 意义
	
	[6] *adv.* 精细地
	
	[7] 摆脱, 除去

- If the wire doesn't go the way you want (it doesn't look the way you want), you can make extra bends in it by clicking in different places on the way (each click will form a corner).

(4) Changing the Name of the Part

- You probably don't want to keep the names C1, C2 etc. , especially if you didn't put the parts in the most logical order. To change the name, double click on the present name (C1, R1 or whatever your part is), then a box will pop up[8] (Edit Reference Designator). In the top window, you can type in the name you want the part to have.

[8] 弹出

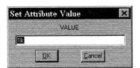

- Please note that if you double click on the part or its value, a different box will appear.

(5) Changing the Value of the Part

- If you only want to change the value of the part (if you don't want all your resistors to be 1kΩ), you can double click on the present value and a box called "Set Attribute Value" will appear. Type in the new value and press "OK". Use μ for micro as in μF = microfarad[9].

[9] n. [电]微法拉

- If you double click on the part itself, you can select "VALUE" and change it in this box.

(6) Making Sure You Have a GND

- This is very important. You cannot do any simulation[10] on the circuit if you don't have a ground. If you aren't sure where to put it, place it near the negative side of your voltage source.

[10] n. 仿真,模拟

(7) Voltage and Current Bubbles
- These are important if you want to measure the voltage at a point or the current going through that point.
- To add voltage or current bubbles[11], go to the right side of the top tool bar and select "Voltage/Level Marker" (Ctrl + M) or "Current Marker". To get either of these, go to "Markers" and either "Voltage/Level Marker" or "Current Marker".

(8) Saving
- To save the circuit, click on the save button on the tool bar (or any other way you normally save files).

(9) Printing
- To print, you must first use your mouse to make a rectangle[12] around your circuit, this is the area of the page that will be printed. Then select print as usual. (You can select).

3. Probe

(1) Before you do the Probe
- You have to have your circuit properly drawn and saved.
- There must not be any floating parts on your page (i.e. unattached[13] devices).
- You should make sure that all parts have the values that you want.
- There are no extra wires.
- It is very important that you have a ground on your circuit.
- Make sure that you have done the Analysis Setup and that only the things you want are enabled.

(2) To Start the Probe
- Click on the Simulate button on the tool bar (or Analysis, Simulate, or F11).
- It will check to make sure you don't have any errors. If you do have errors, correct them.
- Then a new window will pop up. Here is where you can do your graphs.

(3) Graphing
- If you don't have any errors, you should get a window with a black background to pop up (even with errors, it will be OrCAD PSpice A/D Demo).
- If you did have errors, in the bottom, left hand side, it will say what your errors were (these may be difficult to understand, so go to "View - Output File").

(4) Adding/Deleting Traces
- PSpice will automatically put some traces[14] in. You will probably want to change them.
- Go to Trace—Add Trace or on the tool bar. Then select all the traces you want.

[11] *n.* 计划

[12] *n.* 长方形, 矩形

[13] *adj.* 独立的

[14] *n.* 痕迹

- To delete traces, select them on the bottom of the graph and push Delete.

(5) Doing Math

- In Add Traces, there are functions that can be performed, these will add/subtract (or whatever you chose) the lines together.
- Select the first output then either on your keyboard or on the right side, click the function that you wish to perform.
- There are many functions here that may or may not be useful. If you want to know how to use them, you can use PSpice's Help Menu.
- It is interesting to note that you can plot the phase of a value by using IP (xx), where xx is the name of the source you wish to see the phase for.

(6) Labelling

- Click on Text Label ![icon] on top tool bar.
- Type in what you want to write.
- Click "OK".
- You can move this around by single clicking and dragging[15].

(7) Finding Points

- There are Cursor buttons that allow you to find the maximum or minimum or just a point on the line. These are located on the tool bar (to the right).
- Select which curve[16] you want to look at and then select "Toggle Cursor"[17] ![icon].
- Then you can find the max, min, the slope, or the relative max or min (![icon] is find relative max).

(8) Saving

- To save your probe[18] you need to go into the tools menu and click display, this will open up a menu which will allow you to name the probe file and choose where to save it. You can also open previously saved plots from here as well.

(9) Printing

- Select Print in Edit or on the tool bar ![icon].
- Print as usual.

4. Analysis Menu

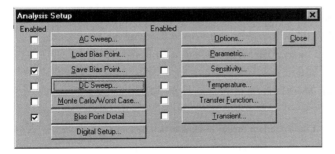

[15] *adj.* 拖曳用的，拖延的

[16] *n.* 曲线
[17] *n.* 指针

[18] *n.* 探示器，取样器

To open the analysis menu click on the button ▣.

(1) AC Sweep

- The AC sweep allows you to plot[19] magnitude versus frequency for different inputs in your circuit.
  In the AC sweep menu you have the choice of three types of analysis: Linear, Octave and Decade.
- These three choices describe the X-axis scaling which will be produced in probe. For example, if you choose decade then a sample of your X-axis might be 10Hz, 1kHz, 100kHz, 10MHz, etc… Therefore if you want to see how your circuit reacts over a very large range of frequencies[20] choose the decade option.
- You now have to specify at how many points you want PSpice to calculate frequencies, and what the start and end frequency will be. That is, over what range of frequencies do you want to simulate your circuit.
- In the AC sweep you also have the option[21] of Noise enable in which PSpice will simulate noise for you either on the output or the input of the circuit. These noise calculations are performed at each frequency step and can be plotted in probe.
- The two types of noise are V(ONOISE) for noise on the outputs and V(INOISE) for noise on the input source.
- To use input noise you need to tell PSpice where you consider the "input" in your circuit to be, for example, if your voltage source is labelled "V1".

(2) DC Sweep

- The DC sweep allows you to do various different sweeps of your circuit to see how it responds to various conditions.
- For all the possible sweeps,
  - voltage,
  - current,
  - temperature, and
  - parameter and global[22].
  you need to specify a start value, an end value, and the number of points you wish to calculate.
- For example, you can sweep your circuit over a voltage range from 0V to 12 V. The main two sweeps that will be most important to us at this stage are the voltage sweep and the current sweep. For these two, you need to indicate to PSpice what component you wish to sweep, for example V1 or V2.
- Another excellent feature of the DC sweep in PSpice, is the ability to do a nested sweep.
- A nested sweep allows you to run two simultaneous sweeps to see how changes in two different DC sources will affect your circuit.
- Once you've filled in the main sweep menu, click on the nested[23] sweep

[19] vt. 画出,标出

[20] n. 频率,周率

[21] n. 选择

[22] adj. 总括的,普遍的

[23] adj. 嵌套的

button and choose the second type of source to sweep and name it, also specifying the start and end values. (Note: In some versions of PSpice you need to click on enable nested sweep). Again you can choose Linear, Octave or Decade, but also you can indicate your own list of values, example: 1V 10V 20V. DO NOT separate the values with commas.

(3) Bias Point Detail

- This is a simple, but incredibly useful sweep. It will not launch Probe and so give you nothing to plot. But by clicking on enable bias[24] current display or enable bias voltage display, this will indicate the voltage and current at certain points within the circuit.

(4) Parametric

- Parametric analysis allows you to run another type of analysis (transient[25], sweeps) while using a range of component values using the global parameter setting. The best way to demonstrate this is with an example, we will use a resistor, but any other standard part would work just as well (capacitor, inductor).
- First, double-click the value label of the resistor that is to be varied. This will open a "Set Attribute Value" dialog box. Enter the name {RVAL} (including the curly[26] braces) in place of the component value. This indicates to PSpice that the value of the resistor is a global parameter called RVAL. In order to define the RVAL parameter in is necessary to place a global parameter list somewhere on the schematic page. To do this, choose "Get New Part" from the menu and select the part named parameter.
- Place the box anywhere on the schematic page. Now double-click on the word PARAMETERS in the box title to bring up the parameter dialog box. Set the NAME1 = value to RVAL (no curly braces) and the VALUE1 = value to the nominal resistance value. This nominal value is required, but it is only used if the DC bias point detail is computed. Otherwise, the value is ignored by PSpice.
- Finally, go to the "Analysis Setup" menu and enable "Parametric[27]" analysis. Open the Parametric setup dialog box and enter the sweep parameters: Name: RVAL Swept variable type: Global Parameter. Make sure the other analysis type(s) are selected in the analysis setup menu (transient, sweeps). PSpice will now automatically perform the simulation over and over, using a new value for RVAL during each run.

(5) Sensitivity

- Sensitivity causes a DC sensitivity analysis to be performed in which one or more output variables may be specified.
- Device sensitivities are provided for the following device types only
  - resistors
  - independent voltage and current sources

[24] n. 偏置，偏压

[25] n. 瞬态

[26] adj. 卷曲的，卷毛的，弯曲的

[27] adj. 参(变)量的

- voltage and current-controlled switches
- diodes
- bipolar[28] transistors

- You would use the sensitivity setting for discovering the maximum range of circuit performance and the causes of extreme operation. These techniques are used to identify effective changes to improve the quality of circuit operation (for example, which components need to have tight tolerance and which can be lower quality and less expensive).
- This isn't as important for us in the lab, but some day when you are constructing real circuits that need to function under various conditions this will be useful.

(6) Temperature

- The temperature option allows you to specify a temperature, or a list of temperatures (do not include commas[29] between temperature values) for which PSpice will simulate your circuit.
- For a list of temperatures that simulation is done for each specified temperature.

(7) Digital Setup

- This paragraph will only indicate the features of the digital setup on the analysis menu, see below for a more complete description.
- In addition to letting you simulate analog circuits, PSpice provides a number of digital parts that can be used in a homogeneous[30] digital circuit, or a heterogeneous analog/digital combination. The digital analysis option allows you to specify the timing of your circuit, by running the gates at their minimum, maximum and typical values. A superb feature allows you to test the worst case timing of your circuit to see how it will operate under these extreme conditions. You also have the option of setting the value of any flip flops you have in your circuit to predefined[31] states which is good to simulate any startup conditions for finite state machines that you are simulating.

(8) Transient

- The transient analysis is probably the most important analysis you can run in PSpice, and it computes various values of your circuit over time. Two very important parameters in the transient analysis are print step and final time.
- The ratio of final time: print step determines how many calculations PSpice must make to plot a wave form. PSpice always defaults the start time to zero second and going until it reaches the user defined final time. It is incredibly important that you think about what print step you should use before running the simulation, if you make the print step too small the probe screen will be cluttered with unnecessary points making it hard to read, and taking extreme amounts of time for PSpice to calculate. However, at the opposite side of that coin is the problem that if you set the print step too high you might miss

[28] adj. 有两极的，双极的

[29] n. 逗号

[30] adj. 同类的，相似的

[31] adj. 预先确定的

important phenomenon that are occurring over very short periods of time in the circuit. Therefore play with step time to see what works best for your circuit. You can set a step ceiling which will limit the size of each interval, thus increasing calculation speed. Another handy[32] feature is the Fourier analysis, which allows you to specify your fundamental frequency and the number of harmonics[33] you wish to see on the plot. PSpice defaults to the 9th harmonic unless you specify otherwise, but this still will allow you to decompose a square wave to see it's components with sufficient detail.	[32] *adj.* 手边的,便利的 [33] *n.* 和声学,谐波

参考译文　可编程逻辑控制器

可编程逻辑控制器(PLCs),也就是可编程控制器,是计算机家族的一员。它们被应用于商业和工业领域。可编程控制器通过监视输入,根据程序做出判决,控制输出来自动控制过程或机器。这一课程意在给你提供基本的 PLC 的功能和配置信息。

1. PLC 的基本操作

PLC 包括输入模块或输入点、一个中央处理单元(CPU)以及输出模块或输出点。输入单元可以接收来自不同现场设备(传感器)的各种数字或模拟信号,然后将它们转换为 CPU 可以使用的逻辑信号。CPU 根据存储器中的程序指令做出判断,然后执行控制指令。输出模块将来自 CPU 的控制指令转换为现场控制的各个设备(执行机构)所能使用的数字或逻辑信号。编程设备用于输入期望的程序指令,由这些指令来决定对于特定的输入,PLC 做什么。操作界面能够显示过程信息及输入新的控制参数。在这个简单的例子里,连接到 PLC 输入端的按钮(传感器),通过电动机的启动器(执行机构)可以启动和停止与 PLC 连接的电动机。

2. 硬接线控制

使用 PLC 之前,许多控制任务是通过接触器和继电器来完成的。这就是经常所说的硬接线控制。硬接线控制必须先设计电路图,确定元件,并且进行安装并建立接线表,电工再连接需要执行特定任务的元件。如果接线中有一个错误,就必须重新连接来纠正。如果改变功能或扩充系统,元件就需要大量变化,并且线路要重接。

3. PLC 的优势

同样或更加复杂的任务,可以用 PLC 来做。设备同继电器触点之间的连线由 PLC 中的程序完成。硬接线虽然依然需要连接到现场设备,但这一要求已很少了。修改应用软件或纠正错误要容易些。在 PLC 中建立或修改程序比起连接或重新连接电路要容易。

下面列出的只是 PLC 的一小部分优点:
- 比通过硬接线方式的物理尺寸小。
- 修改时更简单且更快速。

- PLC 综合了诊断和忽略功能。
- 应用软件能被立即生成文件。
- 应用程序能够被更快并且费用更低地复制。

4. 西门子 PLC

西门子制造了几种属于 SIMATIC S7 产品系列的 PLC。它们是 S7-200、S7-300 和 S7-400 系列。

（1）S7-200

S7-200 由于尺寸小，被称为微型 PLC。S7-200 采用块状设计结构，即它的电源和 I/O 都在一块板上。S7-200 可以被使用在小型的、独立的应用设备中，比如电梯、汽车冲洗器或搅拌机。它也能被使用到更复杂的工业应用中，比如装瓶机和打包机。

（2）S7-300 和 S7-400

S7-300 和 S7-400 型的 PLC 被使用到更复杂的应用中，它们支持更多数量的 I/O 点。这两种 PLC 都是模块化的和可扩展的。由分离模块组成的电源和 I/O 都与 CPU 连接。根据应用的复杂性及未来的扩展需要来选择 S7-300 或 S7-400。西门子销售代表可以提供给你任何型号的西门子 PLC 的附加信息。

5. CPU

CPU 是包括系统存储器的微处理器系统，是 PLC 的决策单元。CPU 监控输入，并根据保存在程序存储器中的指令做出决策。然后 CPU 执行继电、计数、计时、数据比较和顺序操作等功能。

6. 编程设备

程序是在一个编程设备（PG）中建立的，然后被传送到 PLC 中。S7-200 的程序可以使用专用西门子 SIMATIC S7 编程设备来建立，如已经安装了 STEP 7 Micro/WIN 软件，就可以使用 PG 720 或 PG 740。

安装了 STEP 7 Micro/WIN 软件的个人计算机（PC）也可以用作 S7-200 的编程设备。

7. 软件

必须要有一个软件程序告诉 PLC 应服从哪一条指令。编程软件具有 PLC 的特点，一个软件包只针对一个 PLC 或一个 PLC 系列，比如 S7 系列软件包，不能应用到其他 PLC 上。基于 Windows 平台的 S7-200 的编程软件称为 STEP 7-Micro/WIN32。PG 720 和 PG 740 使用前必须先安装 Micro/WIN32 软件。在个人计算机上安装 STEP 7-Micro/WIN32 的方法同安装其他计算机软件的方法类似。

8. 连接电缆 PPI（点对点接口）

连接电缆用来将编程设备的数据传送到 PLC。只有当两个设备使用相同的语言或协议时才能够进行通信。西门子编程设备同 S7-200 之间的通信采用 PPI 协议。对于诸如 PG 720 或 PG 740 的编程设备，需要一个合适的连接电缆。S7-200 使用 9 针的 D 型连接器。

这是一个直插串行接口设备,与西门子编程设备(MPI 接口)兼容,对于其他串行接口它是一个标准连接器。

当个人计算机用作编程设备时也需要一个专用电缆,也就是 PC/PPI 电缆。这根电缆能够使 PLC 的串行口与个人计算机的 RS-232 串行口进行通信。PC/PPI 上的 DIP 开关用来设置 PLC 同个人计算机之间信息传输的合适速度(比特率)。

Unit 9

Text A What Is CNC?

CNC stands for Computer Numerical Control and has been around since the early 1970's. Prior to this, it was called NC, for Numerical Control. (In the early 1970's computers were introduced to these controls, hence the name changed.)

While people in most walks of life have never heard of this term, CNC has touched almost every form of manufacturing process in one way or another. [1] If you'll be working in manufacturing, it's likely that you'll be dealing with CNC on a regular basis. [2]

A drill press can of course be used to machine holes. (It's likely that almost everyone has seen some form of drill press, even if you don't work in manufacturing.) A person can place a drill in the drill chuck that is secured in the spindle of the drill press. They can then (manually) select the desired speed for rotation (commonly by switching belt pulleys), and activate the spindle. Then they manually pull on the quill lever to drive the drill into the workpiece being machined.

As you can easily see, there are a lot of manual interventions required to use a drill press to drill holes. A person is required to do something almost every step along the way! While this manual intervention may be acceptable for manufacturing companies if a small number of holes or workpieces must be machined, as quantities grow, so does the likelihood for fatigue due to the tediousness of the operation. [3] And do note that we've used one of the simplest machining operations (drilling) for our example. There are more complicated machining operations that would require a much higher skill level (and increase the potential for mistakes resulting in scrap workpieces) of the person running the conventional machine tool. (We commonly refer to the style of machine that CNC is replacing as the conventional machine.) [4]

By comparison, the CNC equivalent for a drill press can be programmed to perform this operation in a much more automatic fashion. Everything that the drill press operator was doing manually will now be done by the CNC machine, including: placing the drill in the spindle, activating the spindle, positioning the workpiece under the drill, machining the hole, and turning off the spindle.

1. How CNC works

Here we're relating how CNC works in very general terms.

As you might already have guessed, everything that an operator would be required to do with conventional machine tools is programmable with CNC machines. Once the machine is setup and running, a CNC machine is quite simple to keep running. In fact CNC operators tend to get quite bored during lengthy production runs because there is so little to do. With some CNC machines, even the workpiece loading process has been automated. (We don't mean to over-simplify here. CNC operators are commonly required to do other things related to the CNC operation like measuring workpieces and making adjustments to keep the CNC machine running good workpieces.)

Let's look at some of the specific programmable functions.

2. Motion control

All CNC machine types share this commonality: They all have two or more programmable directions of motion called axes. An axis of motion can be linear or rotary. One of the first specifications that implies a CNC machine's complexity is how many axes it has. Generally speaking, the more axes, the more complex the machine is.

The axes of any CNC machine are required for the purpose of causing the motions needed for the manufacturing process. In the drilling example, these three axes would position the tool over the hole to be machined (in two axes) and machine the hole (with the third axis).[5] Axes are named with letters. Common linear axes names are X, Y, and Z. Common rotary axes names are A, B, and C.

3. Programmable accessories

A CNC machine wouldn't be very helpful if it could only move the workpiece in two or more axes. Almost all CNC machines are programmable in several other ways. The specific CNC machine type has a lot to do with its appropriate programmable accessories. Again, any required function will be programmable on full-blown CNC machine tools.

4. The CNC program

A CNC program is nothing more than another kind of instruction set. It's written in sentence-like format and the control will execute it in sequential order, step by step.

A special series of CNC words are used to communicate what the machine is intended to do. CNC words begin with letter addresses (like F for feedrate, S for spindle speed, and X, Y & Z for axis motion). When placed together in a logical method, a group of CNC words make up a command that resemble a sentence.

For any given CNC machine type, there will only be about 40~50 words used on a regular basis. So if, you compare learning to write CNC programs to learning a foreign language having only 50 words, it shouldn't seem overly difficult to learn CNC programming.[6]

5. The CNC control

The CNC control will interpret a CNC program and activate the series of commands in sequential order. As it reads the program, the CNC control will activate the appropriate machine functions, cause axis motion, and in general, follow the instructions given in the program.

Along with interpreting the CNC program, the CNC control has several other purposes. All current model CNC controls allow programs to be modified (edited) if mistakes are found. The CNC control allows special verification functions (like dry run) to confirm the correctness of the CNC program. The CNC control allows certain important operator inputs to be specified separate from the program, like tool length values. In general, the CNC control has all functions of the machine to be manipulated.

New Words and Phrases

CNC (Computerized Numerical Control)		计算机数值控制
stand for		vt. & vi. 代表,代替,象征
introduce	[ˌɪntrəˈdjuːs]	vt. 传入,引进,提出
walk of life		n. 行业,职业;地位
touch	[tʌtʃ]	n. 触,触觉,接触
		vt. 接触,触摸,触及,达到
		vi. 触摸,接近,涉及,提到
manufacturing	[ˌmænjuˈfæktʃərɪŋ]	n. 制造业
		adj. 制造业的
drill press		n. 钻床
chuck	[tʃʌk]	n. 扔,抛弃
		vt. 用卡盘夹住
spindle	[ˈspɪndl]	n. 主轴,心轴,轴,杆

pulley	['puli]	n. 滑车,滑轮
belt pulley		带轮
quill	[kwil]	n. 钻轴
workpiece	['wə:kpi:s]	n. 工件,加工件
intervention	[,intə(:)'venʃən]	n. 干涉
full-blown		adj. 完善的,成熟的
likelihood	['laiklihud]	n. 可能,可能性
fatigue	[fə'ti:g]	n. 疲乏,疲劳
		vt. 使疲劳
		vi. 疲劳
due to		由于,应归于
tedious	['ti:diəs]	adj. 单调乏味的,沉闷的,冗长乏味的
style	[stail]	n. 类型,式样,字体
bore	[bɔ:]	n. 孔
		vt. & vi. 钻孔
fashion	['fæʃən]	n. 样子,方式,流行,风尚,式样
		vt. 形成,造,制作,把……塑造成,使适应
by comparison		比较
lengthy	['leŋθi]	adj. 过长的
commonality	[,kɔmə'næliti]	n. 普通,常见或平凡的性质或特征
loading	['ləudiŋ]	n. 装载,装填
step by step		一步步,逐步地
axes	['æksi:z]	n. 轴
accessory	[æk'sesəri]	n. 附件,零件,附加物
		adj. 附属的,补充的,副的
feedrate	['fi:dreit]	n. 馈送率,进料速率
overly	['əuvəli]	adv. 过度地,太,非常
interpreting	[in'tə:pritiŋ]	adj. 作为说明的,解释的

Notes

[1] While people in most walks of life have never heard of this term, CNC has touched almost every form of manufacturing process in one way or another.

walks of life 指"各行各业",people in most walks of life 意思是"大部分行业中的人"。in the way 本身表示的是"以……方式",in one way or another 可以理解为"以这样或那样的方式"。

[2] If you'll be working in manufacturing, it's likely that you'll be dealing with CNC on a regular basis.

deal with 表示"打交道,使用"。on...basis 意思是"以……为基准",例如:Rate of work are calculated on weekly basis. 工资是以周为计算基准的。on regular basis 指的就是"以通常或一般为基础"。

〔3〕While this manual intervention may be acceptable for manufacturing companies if a small number of holes or workpieces must be machined, as quantities grow, so does the likelihood for fatigue due to the tediousness of the operation.

本句包含由 while 引导的从句,while 表示"虽然、尽管"。后半句中,so 的意思是"也"。so does the likelihood for fatigue due to the tediousness of the operation 的正常语序是: The likelihood for fatigue grows too, due to the tediousness of the operation. 增长的原因由 due to 引出。

〔4〕There are more complicated machining operations that would require a much higher skill level (and increase the potential for mistakes resulting in scrap workpieces) of the person running the conventional machine tool. (We commonly refer to the style of machine that CNC is replacing as the conventional machine.)

本句中 There are more complicated machining operations 是主句,后面 that 引导的定语从句修饰 operations。在该从句中,running the conventional machine tool 修饰的是 the person。两个括号起进一步说明的作用。第一个括号中的 resulting in 表示前面的行为带来后面的结果,在本句中的意思是,增加了生产废品的出错的可能性。第二个括号中的 that CNC is replacing 是定语从句修饰 the style of machine。

〔5〕In the drilling example, these three axes would position the tool over the hole to be machined (in two axes) and machine the hole (with the third axis).

本句谓语由两部分组成:position the tool 以及 machine the hole。over 表示将 tool 放置在孔的上方,to be machined 不定式表示的是将要被加工的动作。

〔6〕So if you compare learning to write CNC programs to learning a foreign language having only 50 words, it shouldn't seem overly difficult to learn CNC programming.

compare...to...表示"将……比作……",compare...with...表示"用……和……对比",要注意两者的区别。overly 修饰 difficult,表示"非常,极度"。

Exercises

【Ex. 1】 根据课文内容,回答以下问题。

(1) Does CNC play an important role in manufacturing process?

(2) Which machine is easier to operate, conventional machine or CNC machine?

(3) What is the implication of a CNC machine's complexity?

(4) Is it very difficult to learn writing CNC program?

(5) How does the CNC control work?

【Ex. 2】 根据下面的英文解释,写出相应的英文词汇。

英 文 解 释	词　　汇
computerized Numerical Control	
to exercise authoriative or dominating influence over, direct	
a powered vertical drilling machine in which the drill is pressed to the work automatically or by a hand lever	
a system or device, such as a computer, that performs or assists in the performance of a human task	
a simple machine consisting essentially of a wheel with a grooved rim in which a pulled rope or chain can run to change the direction of the pull and thereby lift a load	
a flexible band, as of leather or cloth, worn around the waist to support clothing, secure tools or weapons, or serve as decoration	
one who operates a machine or device	
a reference line from which distances or angles are measured in a coordinate system	
something nonessential but desirable that contributes to an effect or result	
a procedure for solving a problem that involves collection of data, processing, and presentation of results	

【Ex. 3】 把下列句子翻译为中文。

(1) For simple applications (like drilling holes), the CNC program can be developed manually.

(2) In many companies the CAM system will work with the computer aided design (CAD) drawing developed by the company's design engineering department.

(3) The CNC programmer will simply specify the machining operations to be performed and the CAM system will create the CNC program (much like the manual programmer would have written) automatically.

(4) Once the program is developed (either manually or with a CAM system), it must be loaded into the CNC control.

(5) If the program is written manually, it can be typed into any computer using a common word processor (though most companies use a special CNC text editor for this purpose).

(6) The program is in the form of a text file that can be transferred right into the CNC machine. A distributive numerical control (DNC) system is used for this purpose.

(7) A DNC system is nothing more than a computer that is networked with one or more CNC machines.

(8) Regardless of methods, the CNC program must of course be loaded into the CNC machine before it can be run.

(9) Let's look at some of the specific fields and places that emphasis on the manufacturing processes enhanced by CNC machine usage.

(10) Though the setup person could type the program right into the control, this would be like using the CNC machine as a very expensive typewriter.

【Ex. 4】 把下列短文翻译成中文。

As applications get more complicated, and especially when new programs are required on a regular basis, writing programs manually becomes much more difficult. To simplify the programming process, a computer aided manufacturing (CAM) system can be used. A CAM system is a software program that runs on a computer (commonly a PC) that helps the CNC programmer with the programming process. Generally speaking, a CAM system will take the tediousness and drudgery out of programming.

【Ex. 5】 通过 Internet 查找资料,借助如"金山词霸"等电子词典和辅助翻译软件,完成以下技术报告。通过 E-mail 发送给老师,并附你收集资料的网址。

(1) 当前世界上有哪些最主要的数控车床生产厂家以及有哪些最新型的产品(附各种最新产品的图片)。

(2) 通过 Internet 查找用于数控车床的编程语言的基本信息(附上操作界面图片)。

Text B The Basics of Computer Numerical Control

While the specific intention and application for CNC machines vary from one machine type to another, all forms of CNC have common benefits. Though the thrust of this presentation is to teach you CNC usage, it helps to understand why these sophisticated machines have become so popular. Here are a few of but the more important benefits offered by CNC equipment.

The first benefit offered by all forms of CNC machine tools is improved automation. The operator intervention related to producing workpieces can be reduced or eliminated. Many CNC machines can run unattended during their entire machining cycle, freeing the operator to do other tasks. This gives the CNC user several side benefits including reduced operator fatigue, fewer mistakes caused by human error, and consistent and predictable machining time for each workpiece. Since the machine will be running under program control, the skill level required of the CNC operator (related to basic machining practice) is also reduced as compared to a machinist producing workpieces with conventional machine tools.

The second major benefit of CNC technology is consistent and accurate workpieces. Today's CNC machines boast almost unbelievable accuracy and repeatability specifications. This means that once a program is verified, two, ten, or one thousand identical workpieces can be easily produced with precision and consistency.

A third benefit offered by most forms of CNC machine tools is flexibility. Since these machines are run from programs, running a different workpiece is almost as easy as loading a different program. Once a program has been verified and executed for one production run, it can be easily recalled the next time the workpiece is to be ran. This leads to yet another benefit, fast change-overs. Since these machines are very easy to setup and run, and since programs can be easily loaded, they allow very short setup time. This is imperative with today's Just-In-Time product requirements.

1. Motion control—the heart of CNC

The most basic function of any CNC machine is automatic, precise, and consistent motion control. Rather than applying completely mechanical devices to cause motion as is required on most conventional machine tools, CNC machines allow motion control in a revolutionary manner. All forms of CNC equipment have two or more directions of motion, called axes. These axes can be precisely and automatically positioned along their lengths of travel. The two most common axis types are linear (driven along a straight path) and rotary (driven along a circular path).

Instead of causing motion by turning cranks and handwheels as is required on conventional machine tools, CNC machines allow motions to be commanded through programmed commands. Generally speaking, the motion type (rapid, linear, and circular), the axes to move, the amount of motion and the motion rate (feedrate) are programmable with almost all CNC machine tools.

Accurate positioning is accomplished by the operator counting the number of revolutions made on the handwheel plus the graduations on the dial. The drive motor is rotated a corresponding amount, which in turn drives the ball screw, causing linear motion of the axis. A feedback device confirms that the proper amount of ball screw revolutions have occurred.

A CNC command executed within the control (commonly through a program) tells the drive motor to rotate a precise number of times. The rotation of the drive motor in turn rotates the ball screw. And the ball screw causes drives the linear axis. A feedback device at the opposite end of the ball screw allows the control to confirm that the commanded number of rotations has taken place.

Though a rather crude analogy, the same basic linear motion can be found on a common table vise. As you rotate the vise crank, you rotate a lead screw that, in turn, drives the movable jaw on the vise. By comparison, a linear axis on a CNC machine tool is extremely precise. The number of revolutions of the axis drive motor precisely controls the amount of linear motion along the axis.

How axis motion is commanded—understanding coordinate systems. It would be infeasible for the CNC user to cause axis motion by trying to tell each axis drive motor how many times to rotate in order to command a given linear motion amount. (This would be like having to figure out how many turns of the handle on a table vise will cause the movable jaw to move exactly one inch!) Instead, all CNC controls allow axis motion to be commanded in a much simpler and more logical way by utilizing some form of coordinate system. The two most popular coordinate systems used with CNC machines are the rectangular coordinate system and the polar coordinate system. By far, the move popular of these two is the rectangular coordinate system, and we'll use it for all discussions made during this presentation.

One very common application for the rectangular coordinate system is graphing. Almost everyone has had to make or interpret a graph. Since the need to utilize graphs is so commonplace, and since it closely resembles what is required to cause axis motion on a CNC machine, let's review the basis of graphing.

As with any two dimensional graphs, this type of graph has two base lines. Each base line is used to represent something. What the base line represents is broken into increments. Also, each base line has limits. In our productivity example, the horizontal base line is being used to represent time. For this base line, the time increment is in months. Remember this base line has limits—it starts at January and end with December. The vertical base line is representing productivity. Productivity is broken into ten percent increments and starts at zero percent productivity and ends with one hundred percent productivity.

The person making the graph would look up the company's productivity for January of last year and at the productivity position on the graph for January, a point is plotted. This would then be repeated for February, March, and each month of the year. Once all points are plotted, a line or curve can be drawn through each of the points to make it more clear as to how the company did last year.

Let's take what we now know about graphs and relate it to CNC axis motion. Instead of plotting theoretical points to represent conceptual ideas, the CNC programmer is going to be plotting physical end points for axis motions. Each linear axis of the machine tool can be thought of as like a base line of the graph. Like graph base lines, axes are broken into increments. But instead of being broken into increments of conceptual ideas like time and productivity, each linear axis of a CNC machine's rectangular coordinate system is broken into increments of measurement. In the inch mode, the smallest increment is usually 0.0001 inch(in). In the metric mode, the smallest increment is 0.001 millimeter (mm). (By the way, for rotary axes the increment is 0.001°.)

Just like the graph, each axis within the CNC machine's coordinate system must start somewhere. With the graph, the horizontal base line started at January and the vertical base line started at zero percent productivity. This place where the vertical and horizontal base lines come together is called the origin point of the graph. For CNC purposes, this origin point is commonly called the program zero point (also called work zero, part zero, and program origin).

For this example, the two axes we happen to be showing are labelled as X and Y but keep in min that program zero can be applied to any axis. Though the names of each axis will change from one CNC machine type to another (other common names include Z, A, B, C, U, V and W), this example should work nicely to show you how axis motion can be commanded.

The program zero point establishes the point of reference for motion commands in a CNC program. This allows the programmer to specify movements from a common location. If program zero is chosen wisely, usually coordinates needed for the program can be taken directly from the print.

With this technique, if the programmer wishes the tool to be sent to a position one inch to the right of the program zero point, X1.0 is commanded. If the programmer wishes the tool to move to a position one inch above the program zero point, Y1.0 is commanded. The control will automatically determine how many times to rotate each axis drive motor and ball screw to make the axis reach the commanded destination point. This lets the programmer command axis motion in a very logical manner.

With the examples given so far, all points happened to be up and to the right of the program zero point. This area up and to the right of the program zero point is called a quadrant (in this case, quadrant number one). It is not uncommon on CNC machines that end points needed within the program fall in other quadrants. When this happens, at least one of the coordinates must be specified as minus.

2. Understanding absolute versus incremental motion

All discussions to this point assume that the absolute mode of programming is used. The most common CNC word used to designate the absolute mode is G90. In the absolute mode, the end points for all motions will be specified from the program zero point. For beginners, this is usually the best and easiest method of specifying end points for motion commands. However, there is another way of specifying end points for axis motion.

In the incremental mode (commonly specified by G91), end points for motions are specified from the tool's current position, not from program zero. With this method of commanding motion, the programmer must always be asking "How far should I move the tool?" While there are times when the incremental mode can be very helpful. Generally speaking, this is the more cumbersome and difficult method of specifying motion and beginners should concentrate on using the absolute mode.

Be careful when making motion commands. Beginners have the tendency to think incrementally. If working in the absolute mode (as beginners should), the programmer should always be asking "To what position should the tool be moved?" This position is relative to program zero, not from the tools current position.

Aside from making it very easy to determine the current position for any command, another benefit of working in the absolute mode has to do with mistakes made during motion commands. In the absolute mode, if a motion mistake is made in one command of the program, only one movement will be incorrect. On the other hand, if a mistake is made during incremental movements, all motions from the point of the mistake will also be incorrect.

3. Assigning program zero

Keep in mind that the CNC control must be told the location of the program zero point by one means or another. How this is done varies dramatically from one CNC machine and control to another. One (older) method is to assign program zero in the program. With this method, the programmer tells the control how far it is from the program zero point to the starting position of the machine. This is commonly done with a G92 (or G50) command at least at the beginning of the program and possibly at the beginning of each tool.

Another, newer and better way to assign program zero is through some form of offset. Commonly machining center control manufacturers call offsets used to assign program zero fixture offsets. Turning center manufacturers commonly call offsets used to assign program zero for each tool geometry offsets. More on how program zero can be assigned will be presented during key concept number four.

4. Other points about axis motion

To this point, our primary concern has been to show you how to determine the end point of each motion command. As you have seen, doing this requires an understanding of the rectangular coordinate system. However, there are other concerns about how a motion will take place. For example, the type of motion (rapid, straight line, circular, etc.), and motion rate (feedrate), will also be of concern to the programmer. We'll discuss these other considerations during key concept number three.

5. Telling the machine what to do—the CNC program

Almost all current CNC controls use a word address format for programming. (The only exceptions to this are certain conversational controls.) By word address format, we mean that the CNC program is made up of sentence-like commands. Each command is made up of CNC words. Each CNC word has a letter address and a numerical value. The letter address (X, Y, Z, etc.) tells the control the kind of word and the numerical value tells the control the value of the word. Used like words and sentences in the English language, words in a CNC command tell the CNC machine what we wish to do at the present time.

One very good analogy to what happens in a CNC program is found in any set of step by step instructions. For example, you have some visitors coming in from out of town to visit your company. You need to write down instructions to get from the local airport to your company. To do so, you must first be able to visualize the path from the airport to your company. You will then, in sequential order, write down one instruction at a time. The person following your instructions will perform the first step and then go on to the next until he or she reaches your company.

In similar manner, a manual CNC programmer must be able to visualize the machining operations that are to be performed during the execution of the program. Then, in step by step order, the programmer will give a set of commands that makes the machine behave accordingly.

Though slightly off the subject at hand, we wish to make a strong point about visualization. Just as the person developing travel directions MUST be able to visualize the path taken, so MUST the CNC programmer be able to visualize the movements the CNC machine will be making BEFORE a program can be successfully developed. Without this visualization ability, the programmer will not be able to develop the movements in the program correctly. This is one reason why machinists make the best CNC users. An experienced machinist should be able to easily visualize any machining operation taking place.

Just as each concise travel instruction will be made up of one sentence, so will each instruction given within a CNC program be made up of one command. Just as the travel instruction sentence is made up of words (in English), so is the CNC command made up of CNC words (in CNC language).

The person following your set of travel instructions will execute them explicitly. If you make a mistake with your set of instructions, the person will get lost on the way to your company. In similar fashion, the CNC machine will execute a CNC program explicitly. If there is a mistake in the program, the CNC machine will not behave correctly.

Program:

O0001 (Program number)

N005 G54 G90 S400 M03 (Select coordinate system, absolute mode, and turn spindle on CW at 400 RPM)

N010 G00 X1. Y1. (Rapid to XY location of first hole)

N015 G43 H01 Z.1 M08 (Instate tool length compensation, rapid in Z to clearance position above surface to drill, turn on coolant)

N020 G01 Z-1.25 F3.5 (Feed into first hole at 3.5in/min)

N025 G00 Z.1 (Rapid back out of hole) N030 X2. (Rapid to second hole)

N035 G01 Z-1.25 (Feed into the second hole)

N040 G00 Z.1 M09 (Rapid out of the second hole, turn off coolant)

N045 G91 G28 Z0 (Return to reference position in Z)

N050 M30 (End of program command)

While the words and commands in this program probably do not make much sense to you (yet), remember that we are stressing the sequential order by which the CNC program will be executed. The control will first read, interpret and execute the very first command in the program. Only then will it go on to the next command. Read, interpret, execute. Then on to the next command. The control will continue to execute the program in sequential order for the balance of the program. Again, notice the similarity to giving any set of step by step instructions.

6. Other notes about program makeup

As stated programs are made up of commands and commands are made up of word. Each word has a letter address and a numerical value. The letter address tells the control the word type. CNC control manufacturers do vary with regard to how they determine word names (letter addresses) and their meanings. The beginning CNC programmer must reference the control manufacturer's programming manual to determine the word names and meanings. Here is a brief list of some of the word types and their common letter addresses specifications.

O—Program number (Used for program identification)

N—Sequence number (Used for line identification)

G—Preparatory function

X—X axis designation

Y—Y axis designation

Z—Z axis designation

R—Radius designation

F—Feedrate designation

S—Spindle speed designation

H—Tool length offset designation

D—Tool radius offset designation

T—Tool Designation

7. M—Miscellaneous function (See below)

As you can see, many of the letter addresses are chosen in a rather logical manner (T for tool, S for spindle, F for feedrate, etc.). A few require memorizing.

There are two letter addresses (G and M) which allow special functions to be designated. The preparatory function (G) specified is commonly used to set modes. We already introduced absolute mode, specified by G90 and incremental mode, specified by G91. But these are two of the preparatory functions used. You must reference your control manufacturer's manual to find the list of preparatory functions for your particular machine.

Like preparatory functions, miscellaneous functions (M words) allow a variety of special functions. Miscellaneous functions are typically used as programmable switches (like spindle on/off, coolant on/off, and so on). They are also used to allow programming of many other programmable functions of the CNC machine tool.

To a beginner, all of this may seem like CNC programming requires a great deal of memorization. But rest assured that there are only about 30 ~ 40 different words used with CNC programming. If you can think of learning CNC manual programming as learning a foreign language that has only 40 words, it shouldn't seem too difficult.

8. Decimal point programming

Certain letter addresses (CNC words) allow the specification of real numbers (numbers that require portions of a whole number). Examples include X axis designator (X), Y axis designator (Y), and radius designator (R). Almost all current model CNC controls allow a decimal point to be used within the specification of each letter address requiring real numbers. For example, X3.0625 can be used to specify a position along the X axis.

On the other hand, some letter addresses are used to specify integer numbers. Examples include the spindle speed designator (S), the tool station designator (T), sequence numbers (N), preparatory functions (G), and miscellaneous functions (M). For these word types, most controls do NOT allow a decimal point to be used. The beginning programmer must reference the CNC control manufacturer's programming manual to find out which words allow the use of a decimal point.

9. Other programmable functions

All but the very simplest CNC machines have programmable functions other than just axis motion. With today's full blown CNC equipment, almost everything about the machine is programmable. CNC machining centers, for example, allow the spindle speed and direction, coolant, tool changing, and many other functions of the machine to be programmed. In similar fashion, CNC turning centers allow spindle speed and direction, coolant, turret index, and tailstock to be programmed. And all forms of CNC equipment will have their own set of programmable functions. Additionally, certain accessories like probing systems, tool length measuring systems, pallet changers, and adaptive control systems may also be available that require programming considerations.

The list of programmable functions will vary dramatically from one machine to the next, and the user must learn these programmable functions for each CNC machine to be used. In key concept number two, we will take a closer look at what is typically programmable on different forms of CNC machine tools.

New Words and Phrases

intention	[in'tenʃən]	n.	意图,目的
usage	['juːzidʒ]	n.	使用,用法
thrust	[θrʌst]	n.	插
sophisticated	[sə'fistikeitid]	adj.	富有经验的;老练的,练达的;高度发展的,精密复杂的
automation	[ˌɔːtə'meiʃən]	n.	自动控制,自动操作
eliminate	[i'limineit]	vt.	排除,消除
		vt. & vi.	除去
predictable	[pri'diktəb(ə)l]	adj.	可预言的
machinist	[mə'ʃiːnist]	n.	机械师,机械工,机械安装修理工
consistent	[kən'sistənt]	adj.	一致的,调和的,坚固的,相容的
imperative	[im'perətiv]	n.	命令
precise	[pri'sais]	adj.	精确的,准确的
		n.	精确
handwheel	['hændwiːl]	n.	手轮,驾驶盘
graduation	[ˌgrædju'eiʃən]	n.	刻度,分等级
crank	[kræŋk]	n.	曲柄
jaw	[dʒɔː]	n.	叉钳,钳夹
screw	[skruː]	n.	螺钉,螺旋,螺杆,螺孔
		vt.	调节,旋,加强,压榨,强迫,鼓舞
		vi.	转动,旋,拧
vise	[vais]	n.	老虎钳
infeasible	[in'fiːzəbl]	adj.	不可实行的
coordinate	[kəu'ɔːdinit]	n.	同等者,同等物,坐标(用复数)
		adj.	同等的,并列的
		vt.	调整,整理
versus	['vəːsəs]	prep.	对……;与……相对
rectangular	[rek'tæŋgjulə]	adj.	矩形的,成直角的
commonplace	['kɔmənpleis]	n.	平凡的事,平常话
		adj.	平凡的
plot	[plɔt]	vt.	划分,绘图,密谋
		vi.	密谋,策划
conversational	[ˌkɔnvə'seiʃnl]	adj.	对话的,会话的,健谈的
millimeter	['milimiːtə(r)]	n.	毫米(mm)
metric	['metrik]	adj.	米制的,公制的

label	[ˈleibl]	n.	标签,签条,商标,标志
		vt.	贴标签于,指……为,分类,标注
explicit	[iksˈplisit]	adj.	明晰的;明确的,清楚的
miscellaneous	[misiˈleinjəs, - niəs]	adj.	各色各样混在一起
preparatory	[priˈpærətəri]	adj.	预备的
dramatically	[drəˈmætikəli]	adv.	戏剧地,引人注目地

Exercises

【Ex. 6】 根据文章所提供的信息判断正误。

(1) The CNC machine performs operation in a much more automatic fashion.

(2) The advantages of CNC machine include: automation, consistent, accurate workpieces and flexibility.

(3) Although CNC machines have a much more automatic fashion, they require most conventional machine tools to load workpieces.

(4) A CNC instruction executed tells the drive motor to run a approximate number of times.

(5) A feedback device will tell CNC machine if the work will be stopped.

(6) If the CNC user wants to command a linear motion amount, he must tell all axes drive motors how many times to roate.

(7) The CNC machine only uses the rectangular coordiate system.

(8) It is very important to tell the CNC to control the location of the program zero point.

(9) All conversational CNC controls use a word address format for programming.

(10) All CNC control's word names and meanings are the same.

科技英语翻译知识　汉语四字格的运用

　　汉语的四字词组,既包括结构严密、不能随意拆开的四言成语,也包括任意组合而成的四字词组。这种以"四字格"为基本形式的四字词组,言简意赅、形象生动、音节优美、韵律协调。再加上四字格结构灵活多变,它几乎能配置任何一种语法关系,满足结构变化的需要,因此汉语成语有97%采用四字格。在不影响意义表达的情况下,普通的五字词语和三字词语也经常用四字格形式取而代之。英译汉时,四字格若使用恰当,则既能保存原作的风姿,又能使译文大为生色,在进行科技英语翻译时也应该在忠实原文的基础上提倡使用四字格。例如:

(1) Owing to the frequency of collisions between molecules, their motions are entirely at random.

译文一　由于分子之间碰撞频繁,分子运动完全是杂乱的。

译文二　由于分子之间碰撞频繁,分子运动完全是杂乱无章的。

(2) The new computers are indeed cheap and fine.

译文一　这些新计算机确实又便宜又好。

译文二　这些新计算机确实物美价廉。

（3）Copper is an important conductor, both because of its high conductivity and its abundance and low cost.

译文一　铜是一种重要导体，因为它的导电率高，而且数量多，价格低。

译文二　铜是一种重要导体，因为它的导电率高，而且资源丰富，价格低廉。

（4）When we speak, sound waves begin to travel and go in all directions.

译文一　我们说话时，声波便开始向周围扩散。

译文二　我们说话时，声波便开始向四面八方扩散。

（5）In the long run, basic knowledge and technological application go hand in hand——one helps the other.

译文一　从长远来看，基础知识和技术应用是相互一致的。

译文二　从长远来看，基础知识和技术应用是携手并进、相辅相成的。

（6）A large segment of mankind turns to untrammeled nature as a last refuge from encroaching technology.

译文一　许多人都想寻找一块自由的地方，作为他们躲避现代技术侵害的避难所。

译文二　许多人都想寻找一块自由自在的地方，作为他们躲避现代技术侵害的世外桃源。

（7）It is evident that oxygen is the most active element in the atmosphere.

译文一　很显然，氧是大气中最活泼的元素。

译文二　毋庸置疑，氧是大气中最活泼的元素。

（8）All bodies on the earth are known to possess weight.

译文一　大家都知道，地球上的一切物质都有重量。

译文二　众所周知，地球上的一切物质都有重量。

（9）Wireless technology usually suffers when compared to plain-old cable.

译文一　无线电技术和普通电缆相比往往就不利了。

译文二　无线电技术和普通电缆相比往往就相形见绌了。

（10）Electrons and protons carry equal but opposite electrical charges, the number of electrons in the shell of a normal atom is equal to the number of protons in the nucleus.

译文一　电子和质子带有相等但相反的电荷，普通原子外围的电子数与原子核中的质子数相同。

译文二　电子和质子携带数目相等、符号相反的电荷，普通原子外围的电子数正好等于原子核中的质子数。

从以上10个例句可以看出，每个例句的第一个译文虽然都将原文意思表达正确，但是都是平铺直叙，形式上不工整，音节上更谈不上优美，读起来没有回味。而第二个译文，在不改变原意的基础上，由于准确使用四字格，言简意赅、节奏感强、形式工整，既表达了原意，又增加了美感，读起来具有感染力。

但是我们在提倡运用四字格的同时，也要避免乱用四字格，否则会适得其反，以词害义。例如：

This aircraft is small, cheap, pilotless.

译文一 这种飞机小巧玲珑,价廉物美,无人驾驶。

译文二 这种飞机体积不大,价格便宜,无人驾驶。

原文体积上说到小,但没有说"玲珑",价格上说到廉,但没有说"物美"。译文一形式上达到了均匀对称,但是表达的意思超出了原文的范围,因而是不正确的翻译。这时就要宁可舍弃形式上的优美,也要达到词义的准确,如译文二。

总之,四字格的运用是翻译中一个重要的修辞方法。因为四字格可以使译文更加传神达意,增加译文的风采。但是,重要的一点是,四字格运用是以保证原意的传达为基础的,切不可一味追求语言上的美感,而不顾原文的含义乱用四字格,否则会产生相反的效果。

Reading Material

阅读下列文章。

Text	Note

Protel(2)
1. Engineering analysis and verification
  Based on the latest Spice 3f5 and XSpice, Protel DXP's powerful mixed analog and digital circuit simulator[1] operates seamlessly with Protel's schematic editor to effortlessly bring simulation into the design capture process. In Protel DXP you can run mixed-mode[2] circuit simulations directly from your schematic, there's no need to export your design to another application.
  The simulation engine boasts a full complement of standard analyses, as well as advanced temperature and parameter sweeps, and Monte Carlo component tolerance[3] analysis. All simulation parameters are configured through a single intuitive dialog box, and the simulation results are graphically displayed using Protel DXP's integrated waveform viewer, which can display multiple scaled plots simultaneously.
  Protel DXP includes pre-layout and post-layout Signal Integrity Analysis capabilities. Protel DXP's Signal Integrity Analyzer uses sophisticated transmission line calculations, and I/O buffer macro-model information as input for simulations. Based on a fast reflection and crosstalk simulator model, Protel DXP's Signal Integrity Analyzer produces accurate simulations using industry-proven[4] algorithms. Preliminary Impedance and reflection simulations can be run from your source schematics prior to final board layout and routing. This allows you to address potential signal integrity issues such as mismatched[5] net impedances before committing to board layout, to save you time by minimizing board rework.
  Full impedance, signal reflection and crosstalk analysis can be run on your final board to check the real-world performance of your design. Signal integrity screening is build into the Protel DXP design rules system, allowing you to check for gross signal
	[1] *n.* 模拟器
	
	[2] *n.* 混合模型
	
	[3] *n.* 公差
	
	[4] *adj.* 工业检验的
	
	[5] *adj.* 不匹配的

integrity violations as part of the normal board DRC process. When signal integrity issues are found, Protel DXP's Termination Advisor shows you the effects of various termination options, allowing you to settle on the correct solution before modifying your design.

Key Features:
- True Spice 3f5 compliant[6] mixed circuit simulator
- Seamless[7] integration with schematic editor allowing simulation directly from the schematic without netlist export/import
- Digital SimCode language extension to XSpice allows modeling of digital device propagation delays, input and output loading, and power supply-dependent behavior
- Comprehensive analyses, including AC, small signal, transient, noise, and DC Transfer
- Sophisticated component sweep and Monte Carlo analysis modes for testing the effects of component variations and tolerances
- Integrated waveform[8] viewer displays up to 4 scaled plots simultaneously
- Full support for mathematical post-processing of simulation waveforms
- Preliminary[9] impedance and reflection simulations can be run from source schematics prior to final board layout and routing
- Fast simulation of reflection and crosstalk on selected nets with oscilloscope[10]—type display of results, also full zoom and integrated result measurement facilities
- Powerful Termination Advisor for exploring different what—if termination options[11]
- Signal integrity parameters—such as overshoot, undershoot, impedance and signal slope requirements—are specified as standard PCB design rules
- Signal Integrity models are linked into the integrated components

[6] *adj.* 适应的
[7] *adj.* 无缝合线的

[8] *n.* 波形

[9] *adj.* 初步的

[10] *n.* 示波器

[11] *n.* 选择

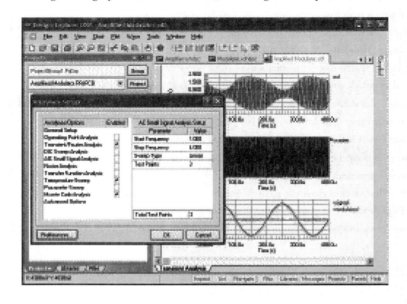

2. Design implementation[12]

Protel DXP's rules-driven PCB layout and editing environment give you full control over the board design process. Define your board using an extensive set of fully-configurable design rules and Protel DXP will enforce relevant[13] rules as you work, minimizing the chance of design errors. Through a powerful hierarchical scoping system you can easily establish rules that exactly specify your design requirements. Instead of using a set of fixed, pre-defined rule scopes, Protel DXP uses queries to define the objects that a rule is applied to. And for even greater flexibility, you can define the order in which rules are applied.

Protel DXP's automatic and interactive component placement features make short work of large placement jobs, and Protel DXP includes enhanced support for multi-channel placement with sophisticated placement room definitions. Protel DXP includes a comprehensive set of intelligent dimensioning[14] tools. Dimension types include: Linear, Datum, Baseline, Leader, Angular, Center, Radial, Linear Diameter, and Radial Diameter. With Protel DXP, dimensions are directly associated with the objects they reference. If an object is moved, the dimensions are automatically updated. You have complete control over the units, precision, text alignment[15], and arrow characteristics for each dimension type.

Protel DXP brings a higher level of control to interactive routing, providing a number of powerful routing modes to suit any routing challenge. While some design tools only highlight rule violations after they occur, Protel DXP monitors relevant rules such as object clearance and track width in real-time[16] as you work— actively preventing violations from occurring and helping you produce error-free boards.

Using technology from Altium's new Situs Autorouting System, Protel DXP brings a new approach to the autorouting challenge—topological[17] path mapping. The autorouter uses topological analysis to find potential[18] routing paths. It then applies powerful routing algorithms to convert the path to a finished route. Unlike traditional shape-based routers, Protel DXP's topological autorouter has the ability to natively find routing paths in non-orthogonal[19] directions, allowing the intelligent assignment of connections to layers. Topological path mapping also allows Protel DXP to efficiently route boards and components of any geometry[20], without the need for extensive post-route cleanup work. The result is high completion rates and a more natural look to the finished board.

Combine this with powerful filtering and selection algorithms, and versatile[21] object and inspection and editing features and you have a board design environment that supports the complex editing processes necessary for successful board design.

Key Features:
- 32 signal, 16 plane & 16 mechanical layers
- Full support for blind and buried vias
- Interactive and automated placement features
- Manual, interactive & automatic routing

[12] n. 执行

[13] adj. 相关的

[14] n. 定尺寸,计算

[15] n. 列队,联结

[16] 实时

[17] adj. 拓扑(学)的
[18] adj. 可能的

[19] adj. 非直角的

[20] n. 几何学

[21] adj. 通用的

- Situs topological autorouter
- Real-time routing rules enforcement
- Supports all component packaging technologies
- Push-&-shove interactive routing
- Complete rules-driven design
- 49 Design rule classes
- Precise targeting[22] of rules
- User-definable rule hierarchy
- Automatic synchronization and updating of source documents and target board design
- True multi-channel design support
- Support for assembly variants
- Reconcile[23] multiple changes in different design documents
- Versatile import and export features:
    - Import OrCAD® layout V9 PCB & library files
    - Import/export AutoCAD® DXF & DWG files up to R14

[22] *n.* 目标,对象

[23] *vt.* 使和谐,使顺从

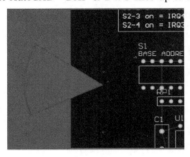

3. Output setup and generation[24]

Protel DXP includes a wide range of output options. These include schematic and PCB printouts[25], netlists, fabrication[26] files, and bills of material.

Protel DXP lets you preview schematic sheets and PCB drawings before printing. From the preview window you can selectively print a range of drawings. You can also copy directly from the preview window to other Windows applications.

Fully integrated into Protel DXP is CAM tastic, Altium's powerful circuit board CAM[27] system that gives you the ultimate in PCB CAM preview and verification. Fabrication files can be viewed directly within Protel DXP using the CAM tastic editor, where you can run an array of fabrication—specific design rule checks.

CAM tastic has been upgraded in Protel DXP and includes a range of new analyses and editing capabilities.

CAM tastic now imports and exports ODB++ files and netlists (IPC-D356), and introduces a new QuickLoad feature that allows you load all supported file types simultaneously.

CAM tastic now includes 18 PCB design check/fix analyses including such as silkscreen[28] over mask, soldier bridging, starved thermals[29] and net antennas.

[24] *n.* 发生
[25] *n.* 打印输出
[26] *n.* 制作,构成

[27] 计算机辅助制造

[28] *n.* 丝网印刷
[29] *n.* 热裂法炭黑

Key Features:
• Project-level definition of output files
   - Output configurations saved with the project
• Supported output types
   - Assembly drawings and pick & place files
   - Schematic & PCB drawings
   - Fabrication files including Gerber, NC Drill and ODB++
   - Netlist output formats include EDIF, VHDL, Spice, Multiwire and Protel
   - Bills of Materials
   - Simulation reports
• Full CAM capabilities
   - Extensive print and inspection tools
   - Import and export to ODB++ or Gerber
   - Fabrication-specific design rule checking
   - NC[30] route path definition
	[30] 数字[值]控制

4. Interfacing to other design tools With Protel DXP you can directly import and export both DXF and native DWG files, providing a strong interface to the mechanical CAD process. Supporting all AutoCAD versions up to release[31] 14, the translation engine supports user-definable layer mapping and allows you to easily relate PCB and AutoCAD layers. There is also full PCB component to AutoCAD block conversion support. Protel DXP includes a wide range of translators to electronic CAD products including OrCAD®, PADS® and P-CAD® design files. Key Features: • Direct import and export of DXF and DWG files • Support for all AutoCAD® versions up to release 14 • Supporting user-definable layer mapping • Ability to easily relate PCB and AutoCAD® layers • Full PCB component to AutoCAD® block support • Direct OrCAD® V9 & V7 schematic and library import • Import PADS® PCB ASCII files • Also import and export P-CAD® PCB Designs, P-CAD® Schematic Designs, and P-CAD® Schematic and PCB Libraries Seamless conversion between Protel DXP and P-CAD® design environments	[31] *vt.* 发行

参考译文　CNC 是什么？

CNC 大约在 1970 年年初就已经出现，用来表示计算机数字控制。在此之前，称为 NC，代表数字控制（在 20 世纪 70 年代初期，计算机被引进到这些控制之中，因此名字改变了）。

虽然大部分人根本没有听说过这个术语，但 CNC 已经以这种或者那种方式涉及制造过程的几乎各个方面。如果你将从事制造业，那你很可能要经常接触 CNC。

钻床当然可以用来钻孔（似乎每个人都见过某种形式的钻床，即使你不从事制造业）。人们将钻头放入钻床的卡盘，卡盘一定要在钻床的轴线上，然后他们（人工地）选择需要的旋转速度（一般是通过切换带轮来实现），再启动钻轴。然后手拉转轴的控制杆使钻头进入被机械加工的工件。

可以容易地看出，在用钻床钻孔的过程中需要许多的人工操作。在整个加工过程中几乎每一步都需要人工参与！虽然加工少量的孔或工件，制造公司可以接受人工操作，但随着数量增加，沉闷的操作可能导致厌烦。并且注意，我们已经用了一个最简单的机械操作（钻孔）作为一个例子。有许多较复杂的机械操作，若要使用传统机械工具，就需要操作人员具有很高的技术水平（并且增加了出错而带来废品的可能）。

通过比较，等同于钻床的 CNC 可以通过执行编制的程序完成操作，这种操作是一种更自动化的方式。钻床操作人员所做的所有人工操作现在都可以由 CNC 机器来完成，包括安装转轴上的钻头、启动主轴、定位钻头下的工件、加工钻孔和停止转轴。

1. CNC 是如何工作的

这里我们将用非常通俗的术语叙述 CNC 是如何工作的。

正如你可能已经猜到的那样，在使用 CNC 机器时利用传统机器工作需要操作者所做的任何事都是可编程的。一旦机器已经设定好并且开始运行，维持 CNC 机器的运行就非常容易了。事实上，由于几乎无事可做，CNC 的操作者在很长的产品加工过程中会感到很无聊。对于有些 CNC 机器，甚至连工件的装卸都是自动的（我们并不是有意过分地简单化。CNC 操作者一般需要做另外一些与 CNC 操作有关的事，比如测量工件，并且进行调整以保证 CNC 机器生产出优质的工件）。

让我们来了解一些特殊的编程功能。

2. 运动控制

所有的 CNC 机器类型都有这个共同点：它们都有两个或更多的称为轴的可编程运动方向。运动轴可能是线性的或旋转的。表明 CNC 机器复杂性最重要的技术指标之一是运动轴的数量。一般来说，轴越多，机器越复杂。

要求 CNC 机器的轴能产生机械加工过程中所需要的运动。在钻孔例子中，需要三个轴定位被加工孔上的工具（沿两个轴）以及加工孔（用第三个轴）。轴用字母命名，一般线性轴用 X、Y 和 Z 表示。一般旋转轴用 A、B 和 C 表示。

3. 编程附件

如果 CNC 机器只能使工件沿两个或更多的轴运动，CNC 机器也不会非常有用。几乎所有的 CNC 机器都可以用几种其他的方式编程。配备了适当可编程附件的特殊类型的 CNC 机器可以做许多事。再者，利用完善的 CNC 机器工具可以对任何需要的功能进行编程。

4. CNC 程序

一个 CNC 程序只是另一种指令集。它以类句格式书写，并且按连续的顺序一步一步地控制执行程序。

一系列特殊的 CNC 字用来传达机器将做什么。CNC 字以字符地址开始（比如，F 代表传送率，S 代表主轴速度，X、Y 和 Z 代表轴运动）。当以逻辑方式组合在一起时，一组 CNC 字就组成类似句子的命令。

对于任何类型的 CNC 机器，通常使用的字只有 40～50 个。如果把学习写 CNC 程序与学习只有 50 个单词的外语进行比较，你不会感到学习 CNC 编程会更加困难。

5. CNC 控制器

CNC 控制器将翻译 CNC 程序，并且按顺序激活一系列命令。在读程序的过程中，CNC 控制器将激活恰当的机器功能，产生轴向运动，一般情况下，按照程序中给出的指令执行。

除了翻译 CNC 程序外，CNC 控制器还有几个其他用途。目前，所有型号的 CNC 控制器都允许发现错误时修改（编辑）程序。CNC 控制器可以进行特别校验（模拟运行）以确保 CNC 程序的正确性。CNC 控制器允许某些将重要操作数据（比如工具长度值）单独输入，独立于程序。一般情况下，CNC 控制器具有人工操作机器的全部功能。

Unit 10

Text A Industrial Bus

1. What is an industrial bus

What is an industrial bus? Traditionally, the industrial bus has been used to allow a central computer to communicate with a field device. The central computer was a mainframe or a MINI (PDP11) and the field device could be a discreet device such as a flow meter, or temperature transmitter or a complex device such as a CNC cell or robot. [1] As the cost of computing power came down, the industrial bus allowed computers to communicate with each other to coordinate industrial production.

As with human languages, many ways were devised to allow the computers and devices to communicate and, as with their human counterpart, most of the communications are incompatible with any of the other systems. [2] The incompatibility can be broken into two categories: the physical layer and the protocol layer.

Two popular industrial buses that use the RS-232 and RS-422/485 standards are Modbus and Data Highway. Modbus was developed by Modicon for its line of PLCs, up to and including the 984 line of controllers. Modbus can be configured for either RS-232 or RS-485 in a 4-wire mode. (Note: Modbus Plus is not RS-232 or RS-485 compatible). Data Highway is the name of the industrial bus. An RS-485 port is available on some PLC-2, 3 and 5 controllers. Consult the manual provided with your controller to be certain of the type of bus supported.

The industrial buses that adhere to the RS-232 and RS-422/485 standard are listed below along with products that are compatible with various industrial buses. B&B products support these buses at the physical layer only and are mainly used as repeaters, line extenders and isolators. B&B also offers a custom design service to solve particular problems that arise from industrial buses.

The physical layer and the protocol layer can be defined using the phone system as an example. Any spoken language can be carried over a phone line. As long as both the speaker and the listener(s) understand the language, communication is possible. The phone line is not concerned with the meaning of the signal that it carries, it is merely moving those signals from one point to another physically. This is the physical layer, the conduit in which communications pass from one point to another. On the other hand, the speaker and listener(s) are concerned with what is transported over the phone line. If the speaker is talking in Spanish and the listener(s) are only fluent in English, communication is not possible. Although the physical layer is working, the language or protocol is not correct, and communications cannot exist. The industrial world has developed a variety of different physical and protocol communications standards. A list of all of them would fill the rest of this article, so we will limit this discussion to industrial buses using the RS-232 and RS-422/485 standards for their physical layer.

The greatest difference between RS-232 and RS-422/485 is the way information is transmitted.

2. Data line isolation theory

When it comes to protect data lines from electrical transients, surge suppression is often the first thing that leaps to mind. The concept of surge suppression is intuitive and there are a large variety of devices on the market to choose from. Models are available to protect everything from your computer to answering machine as well as those serial devices found in RS-232, RS-422 and RS-485 systems. [3]

Unfortunately, in most serial communications systems, surge suppression is not the best choice. The result of most storm and inductively induced surges is to cause a difference in ground potential between points in a communications system. The more physical area covered by the system, the more likely those differences in ground potential will exist.

The water analogy helps explain this. Instead of phenomenon water in a pipe, we'll think a little bigger and use waves on the ocean. Ask anyone what the elevation of the ocean is, and you will get an answer of zero—so common that we call it sea level. While the average ocean elevation is zero, we know that tides and waves can cause large short-term changes in the actual height of the water. This is very similar to earth ground. The effect of a large amount of current dumped into the earth can be visualized in the same way, as a wave propagating outwards from the origin. Until this energy dissipates, the voltage level of the earth will vary greatly between two locations.

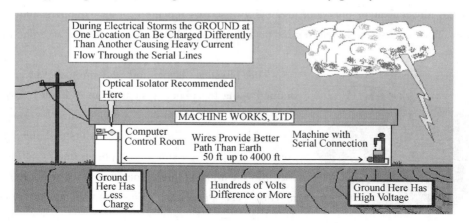

Adding a twist to the ocean analogy, what is the best way to protect a boat from high waves? We could lash the boat to a fixed dock, forcing the boat to remain at one elevation. This will protect against small waves, but this solution obviously has limitations. While a little rough, this comparison isn't far off from what a typical surge suppressor is trying to accomplish.[4] Attempting to clamp a surge of energy to a level safe for the local equipment requires that the clamping device be able to completely absorb or redirect transient energy.[5]

Instead of lashing the boat to a fixed dock let the dock float. Now the boat can rise and fall with the ocean swells (until we hit the end of our floating dock's posts).

Instead of fighting nature, we're simply moving along with it. This is our data line isolation solution.

Isolation is not a new idea. It has always been implemented in telephone and Ethernet equipment. For asynchronous data applications such as many RS-232, RS-422 and RS-485 systems, optical isolators are the most common isolation elements. With isolation, two different grounds (better thought of as reference voltages) can exist on opposite sides of the isolation element without any current flowing through the element.[6] With an optical isolator, this is performed with an LED and a photosensitive transistor. Only light passes between the two elements.

Another benefit of optical isolation is that it is not dependent on installation quality. Typical surge suppressors used in data line protection use special diodes to shunt excess energy to ground. The installer must provide an extremely low impedance ground connection to handle this energy, which can be thousands of amps at frequencies into the tens of megahertz. A small impedance in the ground connection, such as in 1.8m (6ft) of 18 gauge wire, can cause a voltage drop of hundreds of volts—enough voltage to damage most equipments. Isolation, on the other hand, does not require an additional ground connection, making it insensitive to installation quality.

Isolation is not a perfect solution. An additional isolated power supply is required to support the circuitry. This supply may be built in as an isolated DC-DC converter or external. Simple

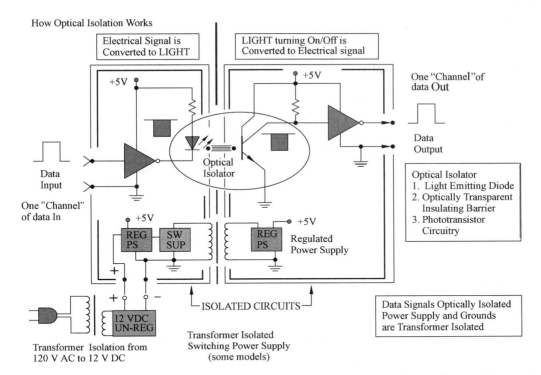

surge suppressors require no power source. Isolation voltages are limited as well, usually ranging from 500V to 4 000V. In some cases, applying both surge suppression and isolation is an effective solution.

When choosing data line protection for a system it is important to consider all available options. There are pros and cons to both surge suppression and optical isolation, however, isolation is a more effective solution for most systems. If in doubt, choose isolation.

New Words and Phrases

mainframe	['mein,freim]	n.	主机,大型机
transmitter	[træns'mitə]	n.	传感器,传送者,变送器,发送器,传递器
counterpart	['kauntəpɑ:t]	n.	副本,极相似的人或物,配对物
incompatible	[,inkəm'pætəbl]	adj.	不相容的,矛盾的
consult	[kən'sʌlt]	vt. & vi.	商量,商议,请教,参考,考虑
conduit	['kɔndit]	n.	导管,水管,水道
leap	[li:p]	vt. & vi.	跳,跳越,跳跃
		n.	跳跃,飞跃
surge	[sə:dʒ]	n.	巨涌,汹涌,澎湃
		vi.	汹涌,澎湃,振荡,滑脱,放松
		vt.	使汹涌奔腾,急放

suppression	[sə'preʃən]	n. 抑制
intuitive	[in'tju(:)itiv]	adj. 直觉的
phenomenon	[fi'nɔminən]	n. 现象
elevation	[ˌeli'veiʃən]	n. 上升,高地,正面图,海拔
dump	[dʌmp]	vt. 倾卸
		n. 堆存处
visualize	['vizjuəlaiz]	vt. 形象,形象化,想象
		vi. 显现
twist	[twist]	n. 一扭,扭曲,螺旋状
		vt. 拧,扭曲,绞,搓,编织
		vi. 扭弯,扭曲,缠绕,扭动
clamp	[klæmp]	n. 夹子,夹具,夹钳
		vt. 夹住,夹紧
redirect	[ˌriːdi'rekt]	vt. (信件)重寄,使改道,使改方向
transient	['trænziənt]	adj. 短暂的,瞬时的
		n. 瞬时现象
isolation	[ˌaisə'leiʃən]	n. 隔绝,孤立,隔离,绝缘
Ethernet	['iːθənet]	n. 以太网
photosensitive	[ˌfəutəu'sensitiv]	adj. 感光性的
optical	['ɔptikəl]	adj. 光学的
megahertz	['megəˌhəːts]	n. 兆赫(MHz)
impedance	[im'piːdəns]	n. 阻抗,全电阻
gauge	[gedʒ]	n. 标准尺,规格,量规,量表
		vt. & vi. 测量
pros	[prəuz]	adv. 正面地
cons	[kɔns]	adv. 反对地,反面
		n. 反对论

Notes

[1] The central computer was a mainframe or a MINI (PDP11) and the field device could be a discreet device such as a flow meter, or temperature transmitter or a complex device such as a CNC cell or robot.

本句说明了两点,一点是 central computer 是什么,一点是 field device 是什么样的设备。在描述 field device 时,举例进行了说明。discreet device 作"智能设备"解释。

[2] As with human languages, many ways were devised to allow the computers and devices to communicate and, as with their human counterpart, most of the communications are incompatible with any of the other systems.

本句分成两部分理解。第一部分将计算机和设备之间的通信和人类语言类比,第二部分将系统兼容情况与人类相比,这两部分内容用 and 连接。其中,with 表示"相对照,相比",

counterpart 表示"相似物"。

[3] Models are available to protect everything from your computer to answering machine as well as those serial devices found in RS-232, RS-422 and RS-485 systems.

本句的基本句型是 models are available to protect everything from…to…as well as…。这样,from…to…就很清楚了,它描写的是 protect 的范围。available 表示"可以得到的,可以利用的"。

[4] While a little rough, this comparison isn't far off from what a typical surge suppressor is trying to accomplish.

本句是包含 while 的转折句,while 表示"尽管"。rough 有很多意思,这里修饰的是 comparison,表示"粗略(的比较)",far off from 表示"远离,远非",本句出现的是 is not far off from,表示的意思正好相反。

[5] Attempting to clamp a surge of energy to a level safe for the local equipment requires that the clamping device be able to completely absorb or redirect transient energy.

本句的主语是动名词短语 attempting to clamp a surge of energy to a level safe for the local equipment,谓语是 require,that 引导宾语从句。动词不定式短语 to clamp a surge of energy to a level safe for the local equipment 是动名词 attempting 的宾语。因为本句的谓语是 require,that 从句里的谓语动词就要用虚拟语气。因为谓语后省略了 should,所以用 be able to,而不是 is able to。

[6] With isolation, two different grounds (better thought of as reference voltages) can exist on opposite sides of the isolation element without any current flowing through the element.

本句中的主干是 two different grounds can exist on opposite sides of the isolation element, with 表示"以……的方式",without 用在动名词前有"不……,未……"的意思,表示前面的动作发生时没有后面的情况出现。

Exercises

【Ex. 1】 根据课文内容,回答以下问题。

(1) What is the traditional use of industrial bus?

(2) What are the factors that lead to communication incompatibility?

(3) What is the first thing to be thought of when it comes to protect data lines from electrical transients?

(4) Why is surge suppression not the best choice in most serial communications systems?

(5) What is the most common isolation element for asynchronous data applications?

【Ex. 2】 根据下面的英文解释,写出相应的英文词汇。

英　文　解　释	词　　汇
a large, powerful computer, often serving several connected terminals	
a copy or duplicate of a legal paper	
a standard procedure for regulating data transmission between computers	
the act of suppressing	
the condition of being elevated	
any of various devices used to join, grip, support, or compress mechanical or structural parts	
remaining in a place only a brief time	
the flexible portion of a whip, such as a plait or thong	
a small electronic device containing a semiconductor and having at least three electrical contacts, used in a circuit as an amplifier, a detector, or a switch	
one million cycles per second. Used especially as a radio-frequency unit	

【Ex. 3】 把下列句子翻译为中文。

(1) All IBM PC and compatible computers are typically equipped with two serial ports and one parallel port. Although these two types of ports are used for communicating with external devices, they work in different ways.

(2) A serial port sends and receives data one bit at a time over one wire. While it takes eight times as long to transfer each byte of data this way, only a few wires are required.

(3) In fact, two-way (full duplex) communications is possible with only three separate wires—one to send, one to receive, and a common signal ground wire.

(4) Some types of serial devices support only one-way communications and therefore use only two wires in the cable—the transmit wire and the signal ground wire.

(5) Both receiver and the transmitter must agree on the number of data bits, as well as the bit rate. Almost all devices transmit data using either 7 or 8 data bits.

(6) A stop bit has a value of 1—or a mark state—and it can be detected correctly even if the previous data bit also had a value of 1.

(7) For example, when even parity is chosen, the parity bit is transmitted with a value of 0 if the number of preceding marks is an even number.

(8) Mark parity means that the parity bit is always set to the mark signal condition and likewise space parity always sends the parity bit in the space signal condition.

(9) DTE stands for Data Terminal Equipment, and DCE stands for Data Communication Equipment. These terms are used to indicate the pin-out for the connectors on a device and the direction of the signals on the pins.

(10) The RS-232 standard states that DTE devices use a 25-pin male connector, and DCE devices use a 25-pin female connector.

【Ex.4】 把下列短文翻译成中文。

Besides the synchronization provided by the use of start and stop bits, an additional bit called a parity bit may optionally be transmitted along with the data. A parity bit affords a small amount of error checking, to help detect data corruption that might occur during transmission. You can choose either even parity, odd parity, mark parity, space parity or none at all. When even or odd parity is being used, the number of marks (logical 1b) in each data byte are counted, and a single bit is transmitted following the data bits to indicate whether the number of 1b just sent is even or odd.

【Ex.5】 通过 Internet 查找资料，借助如"金山词霸"等电子词典和辅助翻译软件，完成以下技术报告。通过 E-mail 发送给老师，并附上收集资料的网址。

(1) 当前工业现场比较流行的总线有哪些？列出它们的主要技术指标和特点。

(2) 当前世界上有哪些最主要的隔离元件生产厂家？有哪些最新型的产品（附上各种最新产品的图片）？

Text B Serial Communications Systems

What do you do when you need to make a dedicated low speed data connection between two places? Need to monitor a PLC on the factory floor from an office area, connect a host PC to a time clock or alarm system, or make a connection between buildings?

The asynchronous serial connection has been the workhorse of low bandwidth communications for decades. For control, monitoring and low volume data transfer asynchronous serial provide a low cost, low development solution. The RS-232 serial port is prevalent on PC's as well as scores of industrial, scientific and consumer devices making it a convenient starting point for communications. Since RS-232 itself is only suited to short connections, many applications require that it be adapted to fit requirements. This article summarizes the choices a system designer has when selecting a serial communications system.

RS-232 or, more currently, EIA-232 uses a single ended, bipolar voltage signal. Voltages typically swing from $-12V$ to $+12V$ with respect to signal ground. Suitable for low noise environments and distances below 30.5m (100ft), RS-232 is commonly used for the desktop modem and mouse. An increase in modem speeds has spurred an effort to increase RS-232 data rate by chip vendors. Transceivers capable of 460kbit and higher are now available, although the actual throughput gains of running higher data rates than 115.2kbit on interrupt based systems is questionable at best.

RS-422 is suited to longer distance communications, up to 1 200m (4 000ft) without repeaters. Using a balanced differential pair results in higher noise immunity than EIA-232. The differential voltage provides a valid signal down to 200mV. Two wires are required for each signal in addition to a signal ground conductor. RS-422 is most commonly used for point-to-point communications, although up to 10 receivers may be connected to a single transmitter.

RS-485 is also suited to longer distance communications, up to 1 200m (4 000ft) without repeaters. Again, a balanced differential pair is used for higher noise immunity than EIA-232. Voltage levels are identical to RS-422. In addition, RS-485 offers a multidrop capability, up to 32 nodes can be connected. The multidrop feature also allows "two-wire" (in addition to signal ground) half-duplex data connection to be made.

Current Loop is the oldest method of connecting serial devices, dating back to Teletype machines. Typically a loop current of 20mA indicates a marking condition and 0mA represents a space. Unfortunately, there is no true standard for current loop, so switching thresholds, voltage requirements and connections vary widely. A well designed current loop system has high noise immunity, and is inherently optically isolated. However, speeds are generally low and the lack of a standard makes connectivity between manufacturers spotty.

Fiber optic communications is growing in popularity as another low bandwidth serial solution.

While costs are still higher than copper solutions, fiber optic links benefit from optimum isolation, noise immunity, and distances up to several miles. Installation of fiber optic cabling requires more care than copper, and repairing damaged cabling is difficult.

RF wireless has become more affordable in recent years and the adoption of spread spectrum technology has further improved performance. Modules to convert RS-232 signals to RF can be used for low to medium data rates. Range is limited, typically several hundred feet, although units are available that reach several miles with appropriate antennas. Higher power units are also available but require an FCC site license to operate. The range and performance of RF wireless is highly depend on the physical and electrical environment and costs are high. If mobility is required or wire isn't possible, wireless has become a viable solution.

COM Type	Pros	Cons
RS-232	Low cost Widely available	Limited distance Poor noise immunity
RS-422	Good noise immunity	
Long distance	May require additional isolation to prevent ground loops	
RS-485	Good noise immunity Long distance Multidrop capability	May require additional isolation to prevent ground loops
Current Loop	High noise immunity Built-in isolation	Low speed Compatibility problems Rarely used in new designs
Fiber Optics	Ideal noise immunity Long distance	More care required in installation Higher initial cost
RF Wireless	High mobility	High cost Sensitive to environmental variables

RS-232 uses a single-ended, bipolar voltage to move data between two points. RS-422/485 uses a balance differential pair to accomplish this same task. The advantage of using RS-422/485 in an industrial environment is greater noise immunity. This allows a greater distance between the transmitter and receiver. There is a downside to the greater distances provided by RS-422/485— the difference of potential between end points.

Industrial buses cover a large area. Often different areas of the network are supplied by different power sources. Even though all of the sources are grounded, a voltage difference can exist between the grounds of these voltage sources. This voltage difference can upset the data line in an RS-422/485 bus by pushing the signal voltage out of range and, in some cases, an excess voltage can damage equipment. Another source of excess voltage potential can be caused by intermittent sources. Power line surges and lightning are causes of this type of disturbance, but other causes, such as large electric motors starting and stopping, can temporarily affect the ground reference voltage. The solution to this problem is to employ RS-422/485 devices that provide isolation between different parts of the network. Additional protection can be achieved by using a fiber optic link between the network and areas known for voltage problems such as a power house or a water treatment plant.

Two popular industrial buses that use the RS-232 and RS-422/485 standards are Modbus and Data Highway. Modbus was developed by Modicon for its line of PLC, up to and including the 984 line of controllers. Modbus can be configured for either RS-232 or RS-485 in a 4-wire mode. (Note: Modbus Plus is not RS-232 or RS-485 compatible). Data Highway is the name of the industrial bus produced by Allen-Bradley and is used on some SLC 500 controllers. An RS-485 port is also available on some PLC-2, PLC-3 and PLC-5 controllers. Consult the manual provided with your controller to be certain of the type of bus supported. The industrial buses that adhere to the RS-232 and RS-422/485 standard are listed below along with products that are compatible with various industrial buses. B&B products support these buses at the physical layer only and are mainly used as repeaters, line extenders and isolators. B&B also offers a custom design service to solve particular problems that arise from industrial buses.

New Words and Phrases

dedicated	['dedikeitid]	adj. 专注的
alarm	[ə'lɑ:m]	n. 警报, 警告器
building	['bildiŋ]	n. 建筑物, 大楼, 工厂
asynchronous	[ei'siŋkrənəs]	adj. 异步的, 不同时的
bandwidth	['bændwidθ]	n. 带宽, 频带宽度
low cost		低成本
convenient	[kən'vi:njənt]	adj. 便利的, 方便的
prevalent	['prevələnt]	adj. 普遍的, 流行的
adapt	[ə'dæpt]	vt. 使适应, 改编
summarize	['sʌməraiz]	vt. & vi. 概述, 总结, 摘要
EIA (Electronic Industries Association)		电子工业协会
modem	['məudem]	n. 调制解调器
swing	[swiŋ]	vt. & vi. 摇摆, 摆动, 回转, 旋转
		n. 摇摆, 摆动
transceiver	[træns'si:və]	n. 无线电收发机, 收发器
questionable	['kwestʃənəb(ə)l]	adj. 有问题的, 不肯定的, 靠不住的
repeater	[ri'pi:tə]	n. 转发器, 中继器
differential	[,difə'renʃəl]	adj. 微分的, 有差别的, 有区别的
		n. 微分
identical	[ai'dentikəl]	adj. 同一的, 同样的
multidrop	['mʌltidrɔp]	n. 多站, 多支路

threshold	[ˈθreʃəuld]	n.	临界值,阈值,门限值,门槛
inherently	[inˈhiərəntli]	adv.	天性地,固有地
fiber = fibre	[ˈfaibə]	n.	光纤
node	[nəud]	n.	节点
half-duplex			半双工
Teletype	[ˈtelitaip]	n.	电传打字机(商标名称)
RF(Radio Freqency)			射频,无线电频率
isolated	[ˈaisəleitid]	adj.	隔离的,孤立的,单独的
connectivity	[kənekˈtiviti]	n.	连通性,连通度
spotty	[ˈspɔti]	adj.	质量不一的
fiber optic			光纤
wireless	[ˈwaiəlis]	adj.	无线的
spectrum	[ˈspektrəm]	n.	光谱,频谱
antenna	[ænˈtenə]	n.	天线
FCC(Federal Communications Commission)			(美国)联邦通信委员会
mobility	[məuˈbiliti]	n.	移动性,流动性
viable	[ˈvaiəbl]	adj.	可行的,实际的,可实施的
accomplish	[əˈkɔmpliʃ]	vt.	完成,达到,实现
industrial environment			工业环境
industrial bus			工业总线
voltage difference			电压差
intermittent	[ˌintə(ː)ˈmitənt]	adj.	间歇的,断断续续的
lightning	[ˈlaitniŋ]	n.	闪电

Exercises

【Ex. 6】 根据文章所提供的信息判断正误。

(1) The asynchronous serial connection is a new technology.
(2) A serial port sends and receives data one bit at a time over one wire.
(3) The serial communication provides a low cost and low development solution.
(4) Almost computers are typically equipped with two RS-422 serial ports.
(5) RS-422 is only suited to short connections.
(6) RS-422 is only suited for low noise environments.
(7) RS-232 is higher noise immunity than RS-422.
(8) A RS-422 transmitter is most commonly used for connection to many receivers.
(9) RS-485 can connect more receivers than RS-422.
(10) The disadvantage of using RS-232 in field is less noise immunity and short connection.

科技英语翻译知识　篇章翻译

以上几个部分谈论的都是有关词汇和单句的翻译。在对这两个部分的翻译有了初步的把握后,应该认识到人们翻译的不会是孤零零的单句,而是完整的语篇材料,或有明确的上下文关系的段落。所以,我们在翻译的过程中,不能从一词、一句甚至一段出发,而应该从整个篇章出发。篇章是一个层次体系,它由段落组成,段落又由句子组成,句子又可分为小句,小句里又有词和词组。词与词、句与句、段落与段落之间都有形式上和内容上的联系。翻译策略体现在每个层次上都有侧重点,体现在段落和篇章层次上。应注意解决句子之间、段落之间的逻辑连接。因此,在翻译过程中,要注意整个语篇的翻译而不是单个句子的简单相加,要以篇章为基础。下面以两个例子做分析。

例一

The best projectiles for this purpose were found to be neutrons. Their mass enables them to pierce the shells of electrons and their small size and electrical neutrality enables them to penetrate the nucleus itself. Once inside the nucleus the intruding neutrons may cause a restructuring and rearrangement of protons and neutrons without causing outward disturbance; in this case, the neutron is absorbed and an isotopic atom of the element is formed. But the intruding neutrons may alternatively disrupt the heavy nucleus, causing it to disintegrate into two or more parts which then become the atoms of two or more different elements; a transmutation of an element has occurred. When heavy atoms are split in this way some loss of mass occurs and this loss of mass is converted into an equivalent quantity of energy according to Einstein's law $E = mc^2$ where c is the velocity of light.

人们发现,用于此目的的发射体就是中子。中子的质量能够粉碎电子壳层,其极小的体积和不带电的特性可以使其自身穿透原子核。一旦进入原子核,"侵入"的中子便可引起质子与中子的重构和重排,且对外层无丝毫影响。此时,中子被吸收,一种新的同位素原子也就诞生了。但是,"侵入"的中子还可能使重原子核分裂,使其分裂成两个或更多部分,成为两种或多种完全不同的元素。原来的元素就发生了蜕变。当重原子以这种方式被分裂时,也就发生了质量损失。根据爱因斯坦的定律 $E = mc^2$(式中 c 为光的速度),这种质量转化成了相等的能量。

分析:英语中为了避免重复,通常用代词指代已提到过的前文事物,代词也因此把两句话连接了起来。原文中第二句话的 their 指代第一句话中出现的"neutrons",因此在翻译时,根据汉语习惯应该重复"中子"。第四句是由两个句子组成的长句子,用分号隔开。前一个分句是说"侵入"的中子可能使重原子核分裂成多个部分,成为不同的元素。后一个分句说原来的元素发生了蜕变。很显然,后一个分句就是前一个分句的结果,因此在翻译中,我们要把这层关系表达出来,而不能孤零零地把第二个分句翻译成"原来的元素发生了蜕变"。

例二

According to growing body of evidence, the chemicals that make up many plastics may migrate out of the material and into foods and fluids, ending up in your body. Once there they could make you very sick indeed. That's what a group of environmental watchdogs has been saying, and the medical community is starting to listen.

一些环保检查人员指出,越来越多的证据表明,许多塑料制品的化学成分会移动到食物或流体上去,最终进入人体内。一旦进入人体,便会使人生病。这一问题也开始受到医学界的关注。

分析:原文第三句"That's what a group of environmental watchdogs has been saying"表示前两句话是谁的观点,所以我们可以根据汉语习惯把它调到第一句。第二句中的there指代的是第一句的in your body,但是在翻译中,要把"人体"重复出来,这样译文才显得自然。listen 是"听"的意思,但是结合整个篇章,我们知道,关于一个问题,人们给予的应该是"关注",因此,这里的listen 做了灵活处理。

Reading Material

阅读下列文章。

Text	Note
VHDL **1. What is VHDL** 　　VHDL is the VHSIC[1] Hardware Description Language. VHSIC is an abbreviation for Very High Speed Integrated Circuits. It can describe the behaviour and structure of electronic systems, but is particularly suited as a language to describe the structure and behaviour of digital electronic hardware designs, such as ASICs[2] and FPGAs[3] as well as conventional digital circuits. 　　VHDL is a notation[4], and is precisely and completely defined by the Language Reference Manual (LRM). This sets VHDL apart from other hardware description languages, which are to some extent defined in an ad hoc way by the behaviour of tools that use them. VHDL is an international standard, regulated by the IEEE[5]. The definition of the language is non-proprietary. 　　VHDL is not an information model, a database schema, a simulator, a toolset or a methodology[6]! However, a methodology and a toolset are essential for the effective use of VHDL.	[1] 超高速集成电路 [2] 特定用途集成电路 [3] 现场可编程门阵列 [4] *n.* 符号 [5] (美国)电气和电子工程师协会 [6] *n.* 方法学,方法论

Simulation and synthesis[7] are the two main kinds of tools which operate on the VHDL language. The Language Reference Manual does not define a simulator, but unambiguously[8] defines what each simulator must do with each part of the language.

VHDL does not constrain the user to one style of description. VHDL allows designs to be described using any methodology—top down, bottom up or middle out! VHDL can be used to describe hardware at the gate level or in a more abstract[9] way. Successful high level design requires a language, a tool set and a suitable methodology. VHDL is the language, you choose the tools, and the methodology.

2. A Brief History of VHDL

(1) The Requirement

The development of VHDL was initiated[10] in 1981 by the United States Department of Defence to address the hardware life cycle crisis. The cost of reprocuring electronic hardware as technologies became obsolete[11] was reaching crisis point, because the function of the parts was not adequately documented, and the various components making up a system were individually verified using a wide range of different and incompatible simulation languages and tools. The requirement was for a language with a wide range of descriptive capability that would work the same on any simulator and was independent of technology or design methodology.

(2) Standardization

The standardization process for VHDL was unique in that the participation[12] and feedback from industry were sought at an early stage. A baseline language (version 7.2) was published 2 years ago before the standard so that tool development could begin in earnest in advance of the standard. All rights to the language definition were given away by the DOD[13] to the IEEE in order to encourage industry acceptance and investment.

(3) ASIC Mandate

DOD Mil[14] Std[15] 454 mandates the supply of a comprehensive VHDL description with every ASIC delivered to the DOD. The best way to provide the required level of description is to use VHDL throughout the design process.

(4) VHDL'93

As an IEEE standard, VHDL must undergo a review process every 5 years (or sooner) to ensure its ongoing relevance to the industry. The first such revision was completed in September 1993, and tools conforming to VHDL'93 are now available. VHDL'98? Hmmm…

3. Levels of Abstraction

VHDL can be used to describe electronic hardware at many different levels of abstraction. When considering the application of VHDL to FPGA/ASIC design, it is helpful to identify and understand the three levels of abstraction shown opposite—algorithm, register transfer level (RTL[16]), and gate level. Algorithms are unsynthesizable, RTL is the input to synthesis, gate level is the output from synthesis. The difference between these levels of abstraction can be understood in

[7] n. 综合,合成

[8] adv. 明白地,不含糊地

[9] adj. 抽象的,深奥的,理论的

[10] vt. & vi. 开始,发起

[11] adj. 陈旧的

[12] n. 分享,参与

[13] (美国)国防部

[14] 军用
[15] 标准

[16] 电阻晶体管逻辑(电路)

terms of timing.

Levels of abstraction in the context of their time domain.

(1) Algorithm

A pure algorithm consists of a set of instructions that are executed in sequence to perform some task. A pure algorithm has neither a clock nor detailed delays. Some aspects of timing can be inferred from the partial ordering of operations within the algorithm. Some synthesis tools (behavioural[17] synthesis) are available that can take algorithmic VHDL code as input. However, even in the case of such tools, the VHDL input may have to be constrained[18] in some artificial[19] way, perhaps through the presence of an "algorithm" clock—operations in the VHDL code can then be synchronized to this clock.

(2) RTL

An RTL description has an explicit clock. All operations are scheduled to occur in specific clock cycles, but there are no detailed delays below the cycle level. Commercially available synthesis tools do allow some freedom in this respect. A single global clock is not required but may be preferred. In addition, retiming is a feature that allows operations to be re-scheduled across clock cycles, though not to the degree permitted in behavioral synthesis tools.

(3) Gates

A gate level description consists of a network of gates and registers instanced from a technology library, which contains technology—specific delay information for each gate.

(4) Writing VHDL for Synthesis

In the diagram above, the RTL level of abstraction is highlighted. This is the ideal level of abstraction at which to design hardware given the state of the art of today's synthesis tools. The gate level is too low a level for describing hardware—remember we're trying to move away from the implementation concerns of hardware design, we want to abstract to the specification level—what the hardware does, not how it does it. Conversely[20], the algorithmic level is too high a level, most commercially available synthesis tools cannot produce hardware from a description at this level.

In the future, as synthesis technology progresses, we will one day view the RTL level of abstraction as the "dirty" way of writing VHDL for hardware and writing algorithmic (often called behavioural) VHDL will be the norm[21].

Until then, VHDL coding at RTL for input to a synthesis tool will give the best results. Getting the best results from your synthesizable RTL VHDL is a key topic of the Comprehensive VHDL and Advanced VHDL Techniques training courses. The latter also covers behavioural synthesis techniques.

4. Scope of VHDL

VHDL is suited to the specification, design and description of digital electronic hardware.

[17] *adj.* 动作的

[18] *adj.* 被强迫的
[19] *adj.* 人造的

[20] *adv.* 倒向地，逆向地，相反地

[21] *n.* 标准，规范

(1) System level

VHDL is not ideally suited for abstract system-level simulation, prior to the hardware-software split. Simulation at this level is usually stochastic[22], and is concerned with modelling performance, throughput, queueing[23] and statistical distributions. VHDL has been used in this area with some success, but is best suited to functional and not stochastic simulation.

(2) Digital

VHDL is suitable for use today in the digital hardware design process, from specification through high-level functional simulation, manual design and logic synthesis down to gate-level simulation. VHDL tools usually provide an integrated design environment in this area.

VHDL is not suited for specialized implementation-level design verification tools such as analog simulation, switch level simulation and worst case timing simulation. VHDL can be used to simulate gate level fan out loading effects providing coding styles are adhered to and delay calculation tools are available. The standardization effort named VITAL (VHDL Initiative Toward ASIC Libraries) is active in this area, and is now bearing fruit in that simulation vendors have built-in VITAL support. More importantly, many ASIC vendors have VITAL-compliant libraries, though not all are allowing VITAL—based sign-off, not yet anyway.

(3) Analogue

Because of VHDL's flexibility as a programming language, it has been stretched[24] to handle analog and switch level simulation in limited cases. However, look out for future standards in the area of analog VHDL.

(4) Design process

The diagram below shows a very simplified view of the electronic system design process incorporating VHDL. The central portion of the diagram shows the parts of the design process which are most impacted[25] by VHDL.

5. Design Flow using VHDL

The diagram below summarizes the high level design flow for an ASIC (i.e. gate array[26], standard cell) or FPGA. In a practical design situation, each step described in the following sections may be split into several smaller steps, and parts of the design flow will be iterated as errors being uncovered.

(1) System-level Verification

As a first step, VHDL may be used to model and simulate aspects of the complete system containing one or more devices. This may be a fully functional description of the system allowing the FPGA/ASIC specification to be validated prior to commencing detailed design. Alternatively, this may be a partial description that abstracts certain properties of the system, such as a performance model to detect system performance bottle-necks[27].

(2) RTL design and testbench creation

Once the overall system architecture and partitioning are stable, the detailed

[22] *adj.* 随机的
[23] *n.* 排队

[24] *vt. & vi.* 伸展, 伸长

[25] *adj.* 压紧的, 结实的

[26] *n.* 排列, 阵列

[27] *n.* 瓶颈

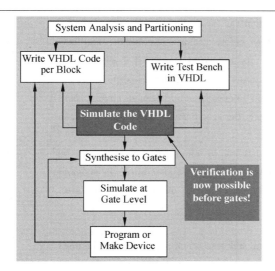

design of each FPGA/ASIC can commence[28]. This starts by capturing the design in VHDL at the register transfer level, and capturing a set of test cases in VHDL. These two tasks are complementary, and are sometimes performed by different design teams in isolation to ensure that the specification is correctly interpreted. The RTL VHDL should be synthesizable if automatic logic synthesis is to be used. Test case generation is a major task that requires a disciplined approach and much engineering ingenuity[29]: the quality of the final FPGA/ASIC depends on the coverage of these test cases.

(3) RTL verification

The RTL VHDL is then simulated to validate the functionality against the specification. RTL simulation is usually one or two orders of magnitude faster than gate level simulation, and experience has shown that this speed-up is best exploited[30] by doing more simulation, not spending less time on simulation.

In practice, it is common to spend 70%～80% of the design cycle writing and simulating VHDL at and above the register transfer level, and 20%～30% of the time synthesizing and verifying the gates.

(4) Look-ahead Synthesis

Although some exploratory synthesis will be done early on in the design process, to provide accurate speed and area data to aid in the evaluation of architectural decisions and to check the engineer's understanding of how the VHDL will be synthesized, the main synthesis production run is deferred until functional simulation is complete. It is pointless to invest a lot of time and effort in synthesis[31] until the functionality of the design is validated.

6. Benefits of using VHDL

(1) Executable specification

It is often reported that a large number of ASIC designs meet their specifications first time, but fail to work when plugged into a system. VHDL allows this issue to be

[28] *vt. & vi.* 开始,着手

[29] *n.* 机灵,独创性,精巧,灵活性

[30] *vt.* 开发

[31] *n.* 综合,合成

addressed in two ways: A VHDL specification can be executed in order to achieve a high level of confidence in its correctness before commencing design, and may simulate one to two orders of magnitude[32] faster than a gate level description. A VHDL specification for a part can form the basis for a simulation model to verify the operation of the part in the wider system context (e.g. printed circuit board simulation). This depends on how accurately the specification handles aspects such as timing and initialization.

Behavioural simulation can reduce design time by allowing design problems to be detected early on, avoiding the need to rework designs at gate level. Behavioural simulation also permits design optimization by exploring alternative architectures, resulting in better designs.

(2) Tools

VHDL descriptions of hardware design and test benches are portable between design tools, and portable between design centres and project partners. You can safely invest in VHDL modelling effort and training, knowing that you will not be tied in to a single tool vendor, but will be free to preserve your investment across tools and platforms. Also, the design automation tool vendors are themselves making a large investment in VHDL, ensuring a continuing supply of state-of-the-art VHDL tools.

(3) Technology

VHDL permits technology independent design through support for top down design and logic synthesis. To move a design to a new technology you need not start from scratch or reverse—engineer a specification—instead you go back up the design tree to a behavioural VHDL description, then implement that in the new technology knowing that the correct functionality will be preserved.

(4) Benefits

- Executable specification
- Validate spec[33] in system context (Subcontract)
- Functionality separated from implementation
- Simulate early and fast (Manage complexity)
- Explore design alternatives
- Get feedback (Produce better designs)
- Automatic synthesis and test generation (ATPG for ASICs)
- Increase productivity (Shorten[34] time-to-market)
- Technology and tool independence (though FPGA features may be unexploited)

[32] *n.* 数量,巨大,广大,量级

[33] *n.* 说明,规格

[34] *vt. & vi.* 缩短,(使)变短

参考译文 工业总线

1. 什么是工业总线

什么是工业总线?传统上来讲,工业总线用于中央计算机同现场设备进行通信。中央

计算机是大型机或小型机,现场设备可能是一个智能设备,如流量表或温度传感器或者是如CNC单元或机器人这样的复杂设备。随着计算设备成本的下降,工业总线允许计算机之间进行相互通信来协调工业生产。

如同人类语言一样,人们用许多方式来设计计算机同设备之间的通信,也同人类相似,大部分通信与任何别的系统之间的通信是不兼容的。不兼容分为两类:物理层和协议层。

Modbus 和 Data Highway 是两种使用 RS-232 和 RS-422/485 标准的流行工业总线。Modbus 是由 Modicon(莫得康)公司为它的 PLC 系列以及 984 系列控制器而开发的。Modbus 可以用 4 类线方式配置 RS-232 或 RS-485(注:Modbus Plus 同 RS-232 或 RS-485 不兼容)。Data Highway 是工业总线的名称。RS-485 接口可以用在 PLC-2、PLC-3 和 PLC-5 等型号的控制器上。参考控制器的随机手册以确定它支持何种总线。使用 RS-232 和 RS-422/485 标准的工业总线以及与多种工业总线都兼容的产品都列在下面。B&B 产品只在物理层支持这些总线,并且主要用做中继器、线性延伸器和隔离器。B&B 也提供定制设计服务,来解决由工业总线产生的特殊问题。

可以用电话系统为例来定义物理层和协议层。任何一种语言都可以通过电话线进行传输。只要对话双方彼此能懂得这种语言,通信就可以实现。电话线不关心传送信号的内容,它只负责在物理上实现将信号从一端传送到另一端。这种将信息从一点传送到另一点的通道就是物理层。另一方面,对话双方关心的是电话线上的传送内容是什么。如果说话者使用西班牙语,而听者只熟悉英语,信息交流是不能实现的。尽管物理层在工作,但语言或者协议如果不正确,信息交流当然无法进行。工业界已经开发了各种不同的物理层和协议层的通信标准。如果将它们全部列出,将用掉这篇文章的所有余下的篇幅。所以,我们这里仅限于讨论将 RS-233 和 RS-422/485 标准作为物理层的工业总线的情况。

RS-232 和 RS-422/485 之间的最大区别在于信息的传送方式不同。

2. 数据线的隔离原理

当要保护数据线免受瞬间的电干扰时,通常最先想到的就是振荡抑制。振荡抑制的概念是直观的,并且市场上有大量不同的振荡抑制设备可供选择。模块可以用来防止从计算机到应答设备以及串行端口为 RS-232、RS-422 或 RS-485 的设备受到任何干扰。

不幸的是,在绝大多数的串行通信系统中,振荡抑制并不是最佳选择。大多数的雷电和感应振荡都会引起通信系统中各点之间对地的电动势差异。系统覆盖的物理区域越大,各点对地电动势的差异就越有可能存在。

和水的类比有助于解释这种现象。我们不要想象管子中的水,而是更大,想象海洋中的浪。问任何人海洋的海拔是多少,你得到的答案是零——因此我们通常称它为海平面。尽管平均海拔是零,我们知道潮汐和海浪都能够引起水面实际高度短暂的很大的变化。这和接地非常相似。大量电流进入大地的效果,同样地,就像水波从原点向四周漫开一样。两点之间的对地电压差距很大,直到这种能量消失。

换个类比角度,在大浪中保护船只最好的方法是什么?我们可以将船只捆在固定的码头上,强制船只保持在某一高度。这种方法能够对付较小的浪,但这种解决方法明显有局限性。尽管这种类比很粗略,但基本描述了典型的振荡抑制器要完成的工作。将局部设备上的能量振荡抑制在一个安全的水平,就需要抑制设备必须能够完全吸收或旁路瞬间的能量。

如果不将船只系在固定的码头上,而是让码头浮动。船只就可以随着巨浪升降(直到碰到浮动码头的柱子的一端)。

不是与自然抗争,而是顺应自然。这是我们的数据线隔离的方法。

隔离不是一个新概念,它早已经被应用在电话和以太网设备上了。对于异步数据传送的应用,比如许多 RS-232、RS-422 和 RS-485 系统,光学隔离器是最常用的隔离元件。通过隔离,两个不同的接地端(最好当作参照电压)处于隔离元件的两边,而不让任何电流穿过隔离元件。对于光学隔离器,是用 LED 和感光的晶体管来完成的,光只在这两个元件之间穿过。

光学隔离的另外一个优点就是它不依赖安装的质量。典型的用于保护数据线的振荡抑制器利用专门的二极管来发散多余的能量到大地。安装者必须提供一个相当低的接地阻抗来处理这些频率可能是数十万兆赫、电流可能是数千安的能量。小的接地阻抗,例如 1.8m (6ft)长,规格为 18 的电线,可以引起数百伏的电压降——这个电压足以损坏大多数设备。而另一方面,对于隔离器,不需要附加接地使得它对安装质量的要求不高。

隔离并不是解决振荡的完美方法。这个回路需要附加的隔离电源。这样提供的电源可能是内置或是外置的一个隔离 DC-DC 转换器。简单的振荡抑制器不需要提供电源。隔离电压也很有限,通常是 500~4 000V。在某种情况下,振荡抑制器和隔离器的组合应用是一个有效的方法。

当为系统选择保护数据线时,考虑全部选项是很重要的。振荡抑制和光学隔离都有优点和缺点,而对于大多数系统来说,隔离是较为有效的方法。如果不能确定,就选择隔离。

自测题

一、根据英文单词,写出中文意思。(20×0.5=10,共10分)

英文单词	汉语意思
load	
terminal	
convert	
resonate	
impedance	
repeater	
effective	
commutator	
velocity	
interaction	
controller	
frequency	
megahertz	
transmitter	
diode	
stray	
amplifier	
capacitance	
ammeter	
oscillation	

二、根据中文意思,写出英文单词。(20×0.5=10,共10分)

汉语意思	英文单词
n. 波形	
n. 打扰,干扰	
n. 千伏特	
n. 反馈,反应	
n. 云母	

adj. 势的,位的

n. 电压,伏特数

n. 损耗

n. 线圈

n. 微伏(等于1伏特的百万分之一)

adj. 可设计的,可编程的

vt. 模拟,模仿

n. 伏特计

adj. 有两极的,双极的

n. 负荷,电荷,费用,充电

n. 电介质,电介体

n. 交流发电机

n. 开关,电闸,转换

n. 逻辑门

n. 半导体

三、根据英文词组,写出中文意思。(30×0.5=15,共15分)

词组	汉译
industrial bus	
voltage difference	
series circuit	
active filter	
inverter switch	
parallel circuit	
discrete input	
positive charge	
alternating current component	
over time	
electric field	
negative charge	
negative direction	
peak value	
equal value resistors	
compound circuits	
conversion of number	
programmable controller	
mechanical energy	
armature field	
right-hand rule	
sampling interval	

analog signal　　　　　　＿＿＿＿＿＿＿＿＿＿＿＿＿＿＿＿＿＿＿＿＿＿
digital signal　　　　　　＿＿＿＿＿＿＿＿＿＿＿＿＿＿＿＿＿＿＿＿＿＿
analog input　　　　　　＿＿＿＿＿＿＿＿＿＿＿＿＿＿＿＿＿＿＿＿＿＿
proximity switch　　　　＿＿＿＿＿＿＿＿＿＿＿＿＿＿＿＿＿＿＿＿＿＿
active power　　　　　　＿＿＿＿＿＿＿＿＿＿＿＿＿＿＿＿＿＿＿＿＿＿
amplifier region　　　　　＿＿＿＿＿＿＿＿＿＿＿＿＿＿＿＿＿＿＿＿＿＿
asynchronous machine　＿＿＿＿＿＿＿＿＿＿＿＿＿＿＿＿＿＿＿＿＿＿
discrete output　　　　　＿＿＿＿＿＿＿＿＿＿＿＿＿＿＿＿＿＿＿＿＿＿

四、根据英文缩写，写出英文完整形式及中文意思。（10×2＝20，共20分）

	英文完整形式	中文意思
RF		
DC		
BCD		
CMOS		
AC		
RPM		
CEMF		
PID		
PLC		
ADC		

五、翻译句子。（10×1.5＝15，共15分）

（1）Ammeters typically include a galvanometer; digital ammeters typically include A/D converters as well.

（2）Generally, an amplifier is a device for increasing the power of a signal.

（3）In electromagnetism and electronics, capacitance is the ability of a body to hold an electrical charge.

（4）All conductors contain electric charges which will move when an electric potential difference (measured in volts) is applied across separate points on the material.

（5）A dielectric is an electrical insulator that can be polarized by an applied electric field.

六、把下列句子翻译为英文。（5×2＝10，共10分）

（1）典型的电感是由导线绕成的线圈。

（2）电流表是一种用来测量电路中电流的测量仪。

（3）交流发电机是机电设备，它把机械能转换成交流电形式的电能。

（4）电容器是一种被动电子元件，它由一对以电介质（绝缘体）分隔的导体组成。

（5）在像铜或铝这样的金属导体中，可移动的带电粒子就是电子。

七、根据下列方框中所给的词填空。（10×1=10，共10分）

（A）vacuum tube （B）forward （C）direction （D）rectification （E）semiconductor
（F）reverse （G）alternating current （H）cathode （I）signals （J）electronic component

In electronics, a diode is a two-terminal ___1___ that conducts electric current in only one ___2___. The term usually refers to a semiconductor diode, the most common type today. This is a crystalline piece of ___3___ material connected to two electrical terminals. A vacuum tube diode (now little used except in some high-power technologies) is a ___4___ with two electrodes: a plate and a ___5___.

The most common function of a diode is to allow an electric current to pass in one direction (called the diode's ___6___ direction), while blocking current in the opposite direction (the ___7___ direction). Thus, the diode can be thought of as an electronic version of a check valve. This unidirectional behavior is called ___8___, and is used to convert ___9___ to direct current, and to extract modulation from radio ___10___ in radio receivers.

八、根据下列短文回答问题，回答请使用英文。（5×2=10，共10分）

Electric Motor

An electric motor converts electrical energy into mechanical energy. Most electric motors operate through interacting magnetic fields and current-carrying conductors to generate force. The reverse process, producing electrical energy from mechanical energy, is done by generators such as an alternator or a dynamo. Many types of electric motors can be run as generators and vice versa. For example, a starter/generator for a gas turbine or traction motors used on vehicles often perform both tasks. Electric motors and generators are commonly referred to as electric machines.

Electric motors are found in applications as diverse as industrial fans, blowers and pumps, machine tools, household appliances, power tools, and disk drives. They may be powered by direct current (e.g., a battery powered portable device or motor vehicle), or by alternating current from a central electrical distribution grid. The smallest motors may be found in electric wristwatches. Medium-size motors of highly standardized dimensions and characteristics provide convenient mechanical power for industrial uses. The very largest electric motors are used for propulsion of ships, pipeline compressors, and water pumps with ratings in the millions of watts. Electric motors may be classified by the source of electric power, by their internal construction, by their application, or by the type of motion they give.

The physical principle of production of mechanical force by the interactions of an electric current and a magnetic field was known as early as 1821. Electric motors of increasing efficiency were constructed throughout the 19th century, but commercial exploitation of electric motors on a large scale required efficient electrical generatorsand electrical distribution networks.

Some devices convert electricity into motion but do not generate usable mechanical power as a primary objective and so are not generally referred to as electric motors. For example, magnetic solenoids and loudspeakers are usually described as actuators and transducers, respectively, instead of motors. On the other hand, some electric motors are simply used as a means of producing torque or force (such as magnetic levitation), but in many cases are still considered to be electric motors, even though they do not actually generate any mechanical energy per se.

(1) What does an electric motor do?
(2) How do most electric motors operate to generate force?
(3) What are the very largest electric motors are used for?
(4) What did commercial exploitation of electric motors on a large scale require?
(5) What are magnetic solenoids and loudspeakers usually described as respectively?

参考答案

一、根据英文单词,写出中文意思。(20×0.5=10,共10分)

英文单词	汉语意思
load	n. 负荷,负载,加载
terminal	n. 终端,接线端,电路接头
convert	n. 转换;vt. 使转变,转换……
resonate	n. 共振,共鸣 vt. 谐振,共鸣,回响,调谐
impedance	n. 阻抗,全电阻
repeater	n. 转发器,中继器
effective	adj. 有效的
commutator	n. 换向器,转接器
velocity	n. 速度,速率,迅速,周转率
interaction	n. 交互作用,交感
controller	n. 控制器
frequency	n. 频率,周率
megahertz	n. 兆赫
transmitter	n. 传感器,变送器,发送器
diode	n. 二极管
stray	n. 杂散电容
amplifier	n. 扩大器
capacitance	n. 容量,电容
ammeter	n. 电表
oscillation	n. 振荡,振动

二、根据中文意思,写出英文单词。(20×0.5=10,共10分)

汉语意思	英文单词
n. 波形	waveform
n. 打扰,干扰	disturbance
n. 千伏特	kilovolt
n. 反馈,反应	feedback

n. 云母	mica
adj. 势的,位的	potential
n. 电压,伏特数	voltage
n. 损耗	depletion
n. 线圈	coil
n. 微伏(等于1伏特的百万分之一)	microvolt
adj. 可设计的,可编程的	programmable
vt. 模拟,模仿	simulate
n. 伏特计	voltmeter
adj. 有两极的,双极的	bipolar
n. 负荷,电荷,费用,充电	charge
n. 电介质,电介体	dielectrics
n. 交流发电机	alternator
n. 开关,电闸,转换	switch
n. 逻辑门	gate
n. 半导体	semiconductor

三、根据英文词组,写出中文意思。(30×0.5=15,共15分)

词 组	汉 译
industrial bus	工业总线
voltage difference	电压差
series circuit	串联电路
active filter	有源滤波器
inverter switch	换向开关
parallel circuit	并联电路
discrete input	开关量输入
positive charge	正电荷
alternating current component	交流分量
over time	超时,滞后
electric field	电场
negative charge	负电荷
negative direction	负向
peak value	峰值
equal value resistors	等值电阻
compound circuits	复合电路
conversion of number	数字转换
programmable controller	可编程控制器
mechanical energy	电能
armature field	电枢磁场

right-hand rule	右手法则
sampling interval	采样间隔
analog signal	模拟信号
digital signal	数字信号
analog input	模拟量输入
proximity switch	接近开关
active power	有功功率
amplifier region	放大区
asynchronous machine	异步电机
discrete output	开关量输出

四、根据英文缩写,写出英文完整形式及中文意思。(10×2=20,共20分)

	英文完整形式	中文意思
RF	Radio Frequency	射频,无线电频率
DC	Direct Current	直流电
BCD	Binary Coded Decimal	二进制编码的十进制
CMOS	Complementary Metal-Oxide-Semi-Conductor Transistor	互补金属氧化物半导体
AC	Alternating Current	交流电
RPM	Revolutions Per Minute	转/分钟
CEMF	Counter Electro Motive Force	反电动势
PID	Proportional Integral Differential	比例、积分、微分
PLC	Programmable Logic Controller	可编程逻辑控制器
ADC	Analog to Digital Converter	模拟/数字转换器

五、翻译句子。(10×1.5=15,共15分)

(1)电流表通常包括一个电流计,而数字电流表通常还包括A/D转换器。

(2)一般来说,放大器是一个提高信号功率的器件。

(3)在电磁学和电子学中,电容是能够保存电量的一个部件。

(4)所有导体都含有电荷,当该材料不同端点间存在电位差时(用伏特计测量)电荷就会移动。

(5)电介质是电绝缘体,可以通过外加电场极化。

六、把下列句子翻译为英文。(5×2=10,共10分)

(1) Typically an inductor is a conducting wire shaped as a coil.

(2) An ammeter is a measuring instrument used to measure the electric current in a circuit.

(3) An alternator is an electromechanical device that converts mechanical energy to electrical energy in the form of alternating current.

(4) A capacitor is a passive electronic component consisting of a pair of conductors separated by a dielectric (insulator).

(5) In metallic conductors, such as copper or aluminum, the movable charged particles are electrons.

七、根据下列方框中所给的词填空。($10 \times 1 = 10$,共 **10** 分)

(1) K (2) C (3) E (4) A (5) I (6) B (7) F (8) D (9) H (10) J

八、根据下列短文回答问题,回答请使用英文。($5 \times 2 = 10$,共 **10** 分)

(1) An electric motor converts electrical energy into mechanical energy.

(2) Most electric motors operate through interacting magnetic fields and current-carrying conductors to generate force.

(3) The very largest electric motors are used for propulsion of ships, pipeline compressors, and water pumps with ratings in the millions of watts.

(4) Commercial exploitation of electric motors on a large scale required efficient electrical generatorsand electrical distribution networks.

(5) Magnetic solenoids and loudspeakers are usually described as actuators and transducers, respectively.

词汇总表

单词表

单　　词	音　　标	意　　义	课次
accommodate	[əˈkɔmədeit]	vt. & vi. 容纳,使……适应	1A
abbreviate	[əˈbriːvieit]	vt. 节略,省略,缩写	2A
abruptly	[əˈbrʌptli]	adv. 突然地,唐突地	7A
absence	[ˈæbsəns]	n. 缺乏,没有	2A
acceleration	[æk,seləˈreiʃən]	n. 加速度	6A
accessory	[ækˈsesəri]	n. 附件,零件,附加物; adj. 附属的,补充的,副的	9A
accomplish	[əˈkɔmpliʃ]	vt. 完成,达到,实现	10B
accomplished	[əˈkɔmpliʃt]	adj. 完成的	6A
accuracy	[ˈækjurəsi]	n. 精确性	1A
accurate	[ˈækjurit]	adj. 准确的,精确的	7B
actuator	[ˈæktjueitə]	n. 操作[执行]机构,执行器	8A
adapt	[əˈdæpt]	vt. 使适应,改编	10B
adjacent	[əˈdʒeisnət]	adj. 邻近的,接近的	4A
advent	[ˈædvənt]	n. (尤指不寻常的人或事)出现,到来	7B
aging	[ˈeidʒiŋ]	n. 老化	
alarm	[əˈlɑːm]	n. 警报,警告器	10B
algebraic	[ˌældʒiˈbreiik]	adj. 代数的,关于代数学的	1B
algorithm	[ˈælgəriðəm]	n. 运算法则	7B
alloy	[ˈælɔi]	n. 合金	
alternator	[ˈɔːltə(ː)neitə]	n. 交流发电机	2A
aluminium	[ˌæljuːˈminjəm]	n. 铝 adj. 铝的	2A
ammeter	[ˈæmitə]	n. 电流表	3A
amp	[æmp]	n. 安培	3B
ampere	[ˈæmpiə]	n. 安培	2A
amplifier	[ˈæmplifaiə]	n. 【电】放大器,扩音器	1B
analogy	[əˈnælədʒi]	n. 模拟,类推	2A
analysis	[əˈnæləsis]	n. 分析,分解	3A
analytical	[ˌænəˈlitikəl]	adj. 解析的	

续表

单词	音标	意义	课次
angular	[ˈæŋgjulə]	adj. 有角的	6A
anode	[ˈænəud]	n. 阳极,正极	
antenna	[ænˈtenə]	n. 天线	10B
apposable	[əˈpəuzəbl]	adj. 并列的	
approach	[əˈprəutʃ]	n. 方法; vt. 接近,动手处理; vi. 靠近	7B
approximation	[əˌprɔksiˈmeiʃən]	n. 近似值,接近,走近	7B
armature	[ˈɑːmətjuə]	n. 电枢(电机的部件)	2B
arsenic	[ˈɑːsənik]	n. [化]砷,砒素	4A
asynchronous	[eiˈsiŋkrənəs]	adj. 异步的,不同时的	10B
atom	[ˈætəm]	n. 原子	2A
attenuate	[əˈtenjueit]	v. 衰减	
attraction	[əˈtrækʃən]	n. 吸引,吸引力	2A
automation	[ɔːtəˈmeiʃən]	n. 自动控制,自动操作	9B
axes	[ˈæksiːz]	n. 轴	9A
bandwidth	[ˈbændwidθ]	n. 带宽,频带宽度	10B
base	[beis]	n. 底部,基础,根据地,基地,本部,基数; vt. 以……作基础,基于……; adj. 低级的	5A
base	[beis]	n. 基极	
battery	[ˈbætəri]	n. 电池	3A
battery	[ˈbætəri]	n. 电池	2A
bearing	[ˈbɛəriŋ]	n. 轴承	6A
behavior	[biˈheivjə]	n. 行为	8B
behaviour	[biˈheivjə]	n. 行为,举止,习性	7A
belt	[belt]	n. 带子	7A
bias	[ˈbaiəs]	n. 偏见,偏爱,斜线; vt. 使存偏见	4B
bimetal	[ˈbaiˈmetl]	n. 双金属材料,双金属器件; adj. 双金属的(= bimetallic)	7A
binary	[ˈbainəri]	adj. 二进制的,二进位的,二元的	5A
bipolar	[baiˈpəulə]	adj. 有两极的,双极的	5B
bit	[bit]	n. 位,比特	5A
blackout	[ˈblækaut]	n. 断电,停电	
bond	[bɔnd]	n. 结合(物),粘结(剂) vt. & vi. 结合	4A
Boolean	[ˈbuːliən]	n. 布尔; adj. 布尔的	5B
boost	[buːst]	v. 增压	
bore	[bɔː]	n. 孔; vt. & vi. 钻孔	9A
bottling	[ˈbɔtliŋ]	n. 装瓶(子)	8A
boundary	[ˈbaundəri]	n. 边界,分界线	4B
branch	[brɑːntʃ]	n. 枝,分支,支流,支脉	3B
breadboard	[ˈbredbɔːd]	n. 电路实验板	8B
breakdown	[ˈbreikdaun]	n. (电)击穿	
breaker	[ˈbreikə]	n. 断路器	

续表

单　　词	音　　标	意　　义	课次
brick	[brik]	n. 砖,砖块,砖形物(如茶砖,冰砖等)	8A
bronze	[brɔnz]	n. 青铜	
brush	[brʌʃ]	n. 电刷	
buck	[bʌk]	n. 补偿	
building	['bildiŋ]	n. 建筑物,大楼,工厂	10B
bus	[bʌs]	n. 母线	
busbar	['bʌsbɑ:]	n. 汇流排,汇流条	2A
bushing	['buʃiŋ]	n. 套管	
button	['bʌtn]	n. 按钮	
bypass	['baipɑ:s]	n. 旁路 vt. 绕过,设旁路,迂回	8B
byte	[bait]	n. (二进制的)字节,位组	5A
calculate	['kælkjuleit]	vt. & vi. 计算,考虑,计划,打算; vt. & vi. (美)以为,认为	3A
calibrate	['kælibreit]	n. 校准	
capability	[,keipə'biliti]	n. 能力,性能,容量	1A
capacitance	[kə'pæsitəns]	n. 容量,电容	1B
capacitive	[kə'pæsitiv]	adj. 电容的	
capacitor	[kə'pæsitə]	n. 电容器(= capacitator)	1B
carrier	['kæriə]	n. 载波	
cathode	['kæθəud]	n. 阴极	
cavity	['kæviti]	n. 空穴,腔	
ceramic	[si'ræmik]	n. 陶瓷,制陶	1B
character	['kæriktə]	n. 字符	8B
characterise	['kærəktəraiz]	vt. 表现……特点,具有……特征 (= characterize)	7A
characteristic	[,kæriktə'ristik]	adj. 特有的,典型的; n. 特性,特征	2A
charge	[tʃɑ:dʒ]	n. 负荷,电荷,费用,充电	2A
chip	[tʃip]	n. 碎片,芯片; vt. & vi. 削成碎片,碎裂	5B
chord	[kɔ:d]	n. 弦	2A
chuck	[tʃʌk]	n. 扔,抛弃; vt. 用卡盘夹住	9A
circuit	['sə:kit]	n. 电路,一圈,周游,巡回	3A
circumvent	[,sə:kəm'vent]	vt. 围绕,包围,智取	5B
clamp	[klæmp]	n. 夹子,夹具,夹钳; vt. 夹住,夹紧	10A
classify	['klæsifai]	vt. 分类,分等	7A
clear	[kliə]	v. 清除(清0)	
closure	['kləuʒə]	n. 关闭; vt. 使终止	8B
coaxial	[kəu'æksəl]	adj. 共轴的,同轴的	
coder	['kəudə]	n. 编码器	
coil	[kɔil]	n. 线圈,卷,圈,盘卷,圈形物,一卷,一圈	6B
collector	[kə'lektə]	n. 集电极	
commercial	[kə'mə:ʃəl]	adj. 商业的,商用的	6A
commonality	[,kɔmə'næliti]	n. 普通,常见或平凡的性质或特征	9A

续表

单　　词	音　　标	意　　义	课次
commonplace	[ˈkɔmənpleis]	n. 平凡的事,平常话；adj. 平凡的	9B
commutator	[ˈkɔmjuteitə]	n. 换向器,转接器	6B
comparator	[ˈkɔmpəreitə]	n. 比较器	
compatible	[kəmˈpætəbl]	adj. 谐调的,一致的,兼容的	8A
compensate	[ˈkɔmpənseit]	vt. & vi. 补偿,偿还	6A
compensation	[kɔmpenˈseiʃən]	n. 补偿	
complementary	[kɔmpləˈmentəri]	adj. 补充的,补足的	5B
complex	[ˈkɔmpleks]	adj. 复杂的,合成的,综合的	8A
complexity	[kəmˈpleksiti]	n. 复杂性	8B
component	[kəmˈpəunənt]	n. 成分	3A
compound	[ˈkɔmpaund]	n. 混合物,[化]化合物；adj. 复合的；vt. & vi. 混合,配合	3B
compound	[ˈkɔmpaund]	v. 复励	
computational	[ˌkɔmpju(ː)ˈteiʃ(ə)n(ə)l]	adj. 计算的	5B
compute	[kəmˈpjuːt]	v. 计算,估计	2A
concentration	[ˌkɔnsenˈtreiʃən]	n. 集中,集合,专心	4B
conductance	[kənˈdʌktəns]	n. 电导,导体,电导系数	2A
conductivity	[ˌkɔndʌkˈtiviti]	n. 传导性,传导率	2A
conductor	[kənˈdʌktə]	n. 导体	2A
conduit	[ˈkɔndit]	n. 导管,水管,水道	10A
configuration	[kənˌfigjuˈreiʃən]	n. 构造,结构,配置,外形	8A
confusion	[kənˈfjuːʒən]	n. 混乱	1A
connectivity	[ˌkɔnekˈtiviti]	n. 连通性,连通度	10B
cons	[kɔns]	adv. 反对地,反面；n. 反对论	10A
consistent	[kənˈsistənt]	adj. 一致的,调和的,坚固的,相容的	9B
constant	[ˈkɔnstənt]	n. 常数,恒量；adj. 不变的,持续的	7A
constant	[ˈkɔnstənt]	n. 常数,恒量 adj. 不变的	2A
consult	[kənˈsʌlt]	vt. & vi. 商量,商议,请教,参考,考虑	10A
contact	[ˈkɔntækt]	n. 接触,触点,联系；vt. 接触,联系	8B
contactor	[ˈkɔntæktə]	n. 接触器	
continuously	[kənˈtinjuəsli]	adv. 不断地,连续地	2B
controller	[kənˈtrəulə]	n. 控制器	7A
convenient	[kənˈviːnjənt]	adj. 便利的,方便的	10B
conventional	[kənˈvenʃnl]	adj. 惯例的,常规的,习俗的,传统的	2A
convergence	[kənˈvədʒəns]	n. 收敛	8B
convergent	[kənˈvəːdʒnt]	adj. 会集于一点的,会聚性的,收敛的	8B
conversational	[ˌkɔnvəˈseiʃnl]	adj. 对话的,会话的,健谈的	9B
convert	[kənˈvəːt]	n. 转换；vt. 使转变,转换……	8A
converter	[kənˈvəːtə]	n. 转换器,变换器	7B
conveyor	[kʌnˈveiə]	n. 传送装置,传送带	6A
cooktop	[ˈkuktɔp]	n. 炉灶	1A

续表

单词	音标	意义	课次
coordinate	[kəuˈɔːdinit]	n. 同等者,同等物,坐标(用复数); adj. 同等的,并列的; vt. 调整,整理	9B
copper	[ˈkɔpə]	n. 铜	1A
corona	[kəˈrəunə]	n. 电晕	
correspond	[kɔrisˈpɔnd]	vi. 符合,协调,相当,相应	7B
corresponding	[ˌkɔrisˈpɔndiŋ]	adj. 相应的	6A
counter	[ˈkauntə]	n. 计数器,计算者	
counterpart	[ˈkauntəpɑːt]	n. 副本,极相似的人或物,配对物	10A
counting	[ˈkauntiŋ]	n. 计数	8A
coupling	[ˈkʌpliŋ]	n. 接合,耦合	8B
covalent	[kəuˈveilənt]	adj. 共有原子价的,共价的	4A
crank	[kræŋk]	n. 曲柄	9B
crystal	[ˈkristl]	n. 晶体,水晶,结晶; adj. 结晶状的	4A
cubic	[ˈkjuːbik]	adj. 立方体的,立方的	2A
current	[ˈkʌrənt]	n. 电流	
curve	[kəːv]	n. 曲线,弯曲 vt. 弯,使弯曲 vi. 成曲形	2A
cycle	[ˈsaik(ə)l]	n. 周期	2B
damp	[dæmp]	vt. 阻尼,使衰减,控制,抑制	
deceleration	[diːseləˈreiʃən]	n. 减速	6A
decimal	[ˈdesiməl]	adj. 十进制,十进的,小数的,以十为基础的; n. 小数	5A
decimal	[ˈdesiməl]	adj. 十进的,小数的,十进制	
decoder	[diːˈkəudə]	n. 解码器,译码器	
dedicated	[ˈdedikeitid]	adj. 专注的	10B
default	[diˈfɔːlt]	n. 默认(值)	8B
deficiency	[diˈfiʃənsi]	n. 缺乏,不足	2A
define	[diˈfain]	vt. 定义,详细说明	6A
density	[ˈdensiti]	n. 密度	5B
depict	[diˈpikt]	vt. 描述,描写	5B
depletion	[diˈpliːʃən]	n. 损耗	4B
derivative	[diˈrivətiv]	n. 导数,微商	7B
derivative	[diˈrivətiv]	adj. 引出的 n. 派生的事物	8B
derive	[diˈraiv]	vt. 得自; vi. 起源	3A
designate	[ˈdezigneit]	vt. 指定(出示),标明	2A
designer	[diˈzainə]	n. 设计者	3A
determine	[diˈtəːmin]	vt. & vi. 决定,确定,测定	2A
deviation	[ˌdiːviˈeiʃən]	n. 偏离,偏向,偏差	7A
diagnostics	[ˌdaiəgˈnɔstiks]	n. 诊断学	8A
diagrammatic	[ˌdaiəgrəˈmætik]	adj. 图表的,概略的	5B
dielectric	[ˌdaiiˈlektrik]	n. 电介质,电介体	1B
differential	[ˌdifəˈrenʃəl]	adj. 微分的,有差别的,有区别的; n. 微分	10B
diffusion	[diˈfjuːʒən]	n. 扩散,传播,漫射	

续表

单词	音标	意义	课次
digit	['didʒit]	n. 阿拉伯数字	5A
dimensional	[di'menʃənəl]	adj. 空间的	4A
diode	['daiəud]	n. 二极管	4A
diode	['daiəud]	n. 二极管	2A
disadvantage	[,disəd'vɑ:ntidʒ]	n. 缺点,不利,不利条件,劣势	7A
discharge	[dis'tʃɑ:dʒ]	n. 放电	
disconnector	[,diskə'nektə]	n. 隔离开关	
discrete	[dis'kri:t]	adj. 不连续的,分散的,离散的	8B
dissipate	['disipeit]	v. 消耗	2A
distance	['distəns]	n. 距离,间隔	6A
distinguish	[dis'tiŋgwiʃ]	vt. & vi. 区别,辨别	2A
distortion	[dis'tɔ:ʃən]	n. 扭曲,变形,失真	
distortion	[dis'tɔ:ʃən]	n. 变形,失真	8B
distribution	[distri'bju:ʃən]	n. 分配,分发	1A
disturbance	[dis'tə:bəns]	n. 打扰,干扰	7A
divider	[di'vaidə]	n. 分割者,间隔物,分配器	3A
domain	[dəu'mein]	n. 范围,领域	7B
drain	[drein]	n. 漏极	
dramatically	[drə'mætikəli]	adv. 戏剧地,引人注目地	9B
drift	[drift]	n. 漂移	
duly	['dju:li]	adv. 适当,合适,适度;当然;及时	7A
dump	[dʌmp]	vt. 倾卸; n. 堆存处	10A
dynamo	['dainəməu]	n. 发电机	
effective	[i'fektiv]	adj. 有效的	2B
electrical	[i'lektrik(ə)l]	adj. 电的,有关电的	2A
electrochemical	[i,lektrəu'kemikəl]	adj. 电气化学的	2A
electromagnet	[ilektrəu'mægnit]	n. 电磁石	6B
electromagnetism	[ilektrəu'mægnitiz(ə)m]	n. 电磁,电磁学	6A
electron	[i'lektrɔn]	n. 电子	2A
elevation	[,eli'veiʃən]	n. 上升,高地,正面图,海拔	10A
elevator	['eliveitə]	n. 电梯,升降机,<空>升降舵	8A
eliminate	[i'limineit]	vt. 排除,消除; vt. & vi. 除去	9B
emitter	[i'mitə]	n. 发射器	1B
emitter	[i'mitə]	n. 发射器,发射极	
energize	['enədʒaiz]	vt. 使活跃,给予精力,加强,给予……电压; vi. 用力,活动	8B
energize	['enədʒaiz]	vt. 给以……电压	
equation	[i'kweiʃən]	n. 相等,平衡,综合体,因素,方程式,等式	3A
equation	[i'kweiʃən]	n. 方程式,等式	7B
equivalent	[i'kwivələnt]	adj. 相等的,相当的,同意义的; n. 等价物,相等物	3B

续表

单　词	音　标	意　义	课次
equivalent	[iˈkwivələnt]	adj. 相等的,相当的,同意义的; n. 等价物,相等物	5A
equivalent	[iˈkwivələnt]	adj. 相等的,相当的; n. 等价物,相等物	6A
estimate	[ˈestimeit]	vt. & vi. 估计,估价,评估; n. 估计,估价,评估	7A
exact	[igˈzækt]	adj. 精确的,准确的	2A
exactly	[igˈzæktli]	adv. 精确地,确切地	7B
excess	[ikˈses]	n. 过度,多余,超过,超额; adj. 过度的,额外的	2A
excitation	[ˌeksiˈteiʃən]	n. 励磁	
exciter	[ikˈsaitə]	n. 励磁器	
experimentation	[eksˌperimenˈteiʃən]	n. 实验,实验法	4A
explanatory	[iksˈplænətəri]	adj. 说明的,解释性的	1A
explicit	[iksˈplisit]	adj. 明晰的;明确的,清楚的	9B
exploit	[iksˈplɔit]	vt. 开拓,开发,开采	4B
exponent	[eksˈpəunənt]	n. 指数,幂	2A
express	[ikˈspres]	vt. 表达,表示	6A
extensive	[iksˈtensiv]	adj. 广大的,广阔的,广泛的	8A
fabricate	[ˈfæbrikeit]	vt. 制作,构成	4A
facility	[fəˈsiliti]	n. 容易,熟练,便利,敏捷,设备,工具	7A
factor	[ˈfæktə]	n. 因素,要素	7A
Farad	[ˈfærəd]	n. 法拉	
fashion	[ˈfæʃən]	n. 样子,方式,流行,风尚,式样; vt. 形成,造,制作,把……塑造成,使适应	9A
fatigue	[fəˈti:g]	n. 疲乏,疲劳; vt. 使疲劳; vi. 疲劳	9A
fault	[fɔ:lt]	n. 故障	
feedback	[ˈfi:dbæk]	n. 反馈,反应	7A
feeder	[ˈfi:də]	n. 馈电线	
feedrate	[ˈfi:dreit]	n. 馈送率,进料速率	9A
fiber = fibre	[ˈfaibə]	n. 光纤	10B
field	[fi:ld]	n. 原野,现场	8A
filter	[ˈfiltə]	n. 滤波器	
flashover	[ˈflæʃˌəuvə]	n. 闪络,跳火	
flexibility	[ˌfleksəˈbiliti]	n. 弹性,适应性,机动性,挠性	7B
flux	[flʌks]	n. 流量,通量; vi. 熔化,流出; vt. 使熔融	6B
formidable	[ˈfɔ:midəbl]	adj. 强大的,令人敬畏的,艰难的	5B
formula	[ˈfɔ:mjulə]	n. 公式,规则	2A
frequency	[ˈfri:kwənsi]	n. 频率,周率	2B
friction	[ˈfrikʃən]	n. 摩擦,摩擦力	6A
fundamental	[ˌfʌndəˈmentl]	adj. 基础的,基本的	3A
fuse	[fju:z]	n. 熔断器,保险丝,熔丝	
gain	[gein]	n. &v. 增益	

续表

单 词	音 标	意 义	课次
gallium	['gæliəm]	n. 镓	4A
gate	[geit]	n. 逻辑门,门电路	5B
gauge	[gedʒ]	n. 标准尺,规格,量规,量表;vt. & vi. 测量	10A
generalize	['dʒenərəlaiz]	vt. 归纳,概括,推广,普及	7B
generate	['dʒenə,reit]	vt. 发电	
generator	['dʒenəreitə]	n. 发电机	2A
geometry	[dʒi'ɔmitri]	n. 几何形状	2A
germanium	[dʒə:'meiniəm]	n. 锗	4A
gotcha	['gɔtʃə]	<口> = (I have) got you	1B
graduation	[,grædju'eiʃən]	n. 刻度,分等级	9B
grap	[grɑ:f]	n. 图表,曲线图	8B
gravity	['græviti]	n. 地心引力,重力	2A
grid	[grid]	n. 栅极	
grounding	['graundiŋ]	n. 接地	
half-adder	[hɑ:f-'ædə]	n. 半加器	
halve	[hɑ:v]	vt. 二等分,平分	1A
handle	['hændl]	vt. 触摸,运用,买卖,处理,操作; vi. 搬运,易于操纵	8A
handwheel	['hændwi:l]	n. 手轮,驾驶盘	9B
hardware	['hɑ:dwɛə]	n. (计算机的)硬件,(电子仪器的)部件	7B
harmonic	[hɑ:'mɔnik]	n. 谐波	
havoc	['hævək]	n. 严重破坏;vt. 损害	1B
hence	[hens]	adv. 因此,从此	7B
Henry	['henri]	n. 亨利	
Hertz	['hə:ts]	n. 赫(Hz),赫兹	2B
hexadecimal	[heksə'desim(ə)l]	adj. 十六进制的;n. 十六进制	5A
high-gain	[hai-gein]	n. 高增益	
hole	[həul]	n. 空穴	
horizontal	[,hɔri'zɔntl]	adj. 水平的	2B
horsepower	['hɔ:s,pauə]	n. 马力	6A
hydraulics	['hai'drɔ:liks]	n. 水力学	2A
identical	[ai'dentikəl]	adj. 同一的,同样的	10B
identical	[ai'dentikəl]	adj. 同一的,同样的	2A
ignore	[ig'nɔ:]	vt. 忽略,不理睬,忽视	2A
illuminate	[i'lju:mineit]	vt. 阐明,说明(问题等),启发,启蒙	3A
imbalance	[im'bæləns]	n. 不平衡,不均衡	4B
immunity	[i'mju:niti]	n. 免疫性	5B
impedance	[im'pi:dəns]	n. 阻抗,全电阻	10A
imperative	[im'perətiv]	n. 命令	9B
implement	['implimənt]	n. 工具,器具;vt. 贯彻,实现;vt. & vi. 执行	7B

续表

单词	音标	意义	课次
impractical	[im'præktikəl]	adj. 不切实际的,昧于实际的	8B
impurity	[im'pjuəriti]	n. 杂质,混杂物,不洁,不纯	4A
inadequate	[in'ædikwit]	adj. 不充分的,不适当的	8B
incident	['insidənt]	adj. 入射的	
incompatible	[,inkəm'pætəbl]	adj. 不相容的,矛盾的	10A
inductance	[in'dʌktəns]	n. 电感(量),感应系数	
induction	[in'dʌkʃən]	n. 感应	
inductive	[in'dʌktiv]	adj. 电感的	
inductor	[in'dʌktə]	n. 电感器,感应器	1B
inertia	[i'nə:ʃjə]	n. 惯性,惯量	6A
infeasible	[in'fi:zəbl]	adj. 不可实行的	9B
inherent	[in'hiərənt]	adj. 固有的,内在的,与生俱来的	1B
inherently	[in'hiərəntli]	adv. 天性地,固有地	10B
inhibit	[in'hibit]	v. 抑制,约束	2A
initialization	[i,niʃəlai'zeiʃən]	n. 设定初值,初始化	8B
instantaneous	[,instən'teinjəs]	adj. 瞬间的,即刻的,即时的	2B
instrument	['instrumənt]	n. 工具,器械,器具	2A
insulation	[,insju'leiʃən]	n. 绝缘	
insulator	['insjuleitə]	n. 绝缘体,绝热器	2A
insulator	['insjuleitə]	n. 绝缘体	1A
integral	['intigrəl]	adj. 积分的;n. 积分	7B
integration	[,inti'greiʃən]	n. 综合	5B
integrity	[in'tegriti]	n. 完整性	8B
intention	[in'tenʃən]	n. 意图,目的	9B
interaction	[,intər'ækʃən]	n. 交互作用,交感	4B
interface	['intə(:),feis]	n. 分界面,<计>界面,接口	8A
intermittent	[,intə(:)'mitənt]	adj. 间歇的,断断续续的	10B
interpreting	[in'tə:pritiŋ]	adj. 作为说明的,解释的	9A
intervention	[,intə(:)'venʃən]	n. 干涉	9A
intrinsic	[in'trinsik]	adj. (指价值、性质)固有的,内在的,本质的	4A
introduce	[,intrə'dju:s]	vt. 传入,引进,提出	9A
intuitive	[in'tju(:)itiv]	adj. 直觉的	10A
inverse	['in'və:s]	adv. 相反地,倒转地	3A
invert	[in'və:t]	adj. 转化的;vt. 使颠倒,使转化;n. 颠倒的事物	3B
inverter	[in'və:tə]	n. 反用换流器,变极器	5B
isolated	['aisəleitid]	adj. 隔离的,孤立的,单独的	10B
isolation	[,aisə'leiʃən]	n. 隔绝,孤立,隔离,绝缘	10A
isolator	['aisəleitə]	n. 刀闸(隔离开关),绝缘体	
jaw	[dʒɔ:]	n. 叉钳,钳夹	9B
Joule	[dʒu:l]	n. 焦耳	

续表

单　词	音　标	意　义	课次
jug	[dʒʌg]	n. 水壶	1A
junction	[ˈdʒʌŋkʃən]	n. 连接,接合,交叉点,汇合处	4B
kilovolt	[ˈkiləuvəult]	n. 千伏特	2A
knob	[nɔb]	n. (门,抽屉等的)球形把手,节,旋钮	7A
label	[ˈleibl]	n. 标签,签条,商标,标志; vt. 贴标签于,指……为,分类,标注	9B
lamination	[ˌlæmiˈneiʃən]	n. 叠片	
lamp	[læmp]	n. 灯	3A
laptop	[ˈlæptɔp]	n. 便携式电脑,膝上型电脑	5B
lateral	[ˈlætərəl]	n. 支线	
leakage	[ˈliːkidʒ]	n. 漏,泄漏,渗漏	4B
leap	[liːp]	vt. & vi. 跳,跳越,跳跃; n. 跳跃,飞跃	10A
lengthy	[ˈleŋθi]	adj. 过长的	9A
lever	[ˈliːvə, ˈlevə]	n. 电平,杆,控制杆; vt. & vi. 抬起	6A
lightning	[ˈlaitniŋ]	n. 闪电	10B
likelihood	[ˈlaiklihud]	n. 可能,可能性	9A
limiter	[ˈlimitə]	n. 限幅器	
linear	[ˈliniə]	adj. 线的,直线的,线性的	6A
linearize	[ˈliniəˌraiz]	vt. 使线性化	8B
load	[ləud]	n. 负荷,负载,加载	3A
loading	[ˈləudiŋ]	n. 装载,装填	9A
loop	[luːp]	n. 回路	
loss	[lɔs]	n. 损耗	6A
machinist	[məˈʃiːnist]	n. 机械师,机械工,机械安装修理工	9B
magnet	[ˈmægnit]	n. 磁体,磁铁	6B
magnetic	[mægˈnetic]	adj. 磁的,有磁性的,有吸引力的	2A
magnitude	[ˈmægnitjuːd]	n. 大小,数量,幅度	2B
mainframe	[ˈmeinfreim]	n. 主机,大型机	10A
malfunction	[mælˈfʌŋkʃən]	n. 故障,失灵	
manipulate	[məˈnipjuleit]	vt. (熟练地)操作,使用(机器等)	7A
manufacturing	[ˌmænjuˈfæktʃəriŋ]	n. 制造业; adj. 制造业的	9A
margin	[ˈmɑːdʒin]	n. 裕度	
material	[məˈtiəriəl]	n. 材料,原料	1A
mathematician	[ˌmæθiməˈtiʃən]	n. 数学家	3A
maximum	[ˈmæksiməm]	n. 最大量,最大值,极大,极大值	2B
measure	[ˈmeʒə]	n. 测量,量度器,量度标准; vt. 测量,测度,估量,调节	6A
measurement	[ˈmeʒəmənt]	n. 量度,测量法	2A
megahertz	[ˈmegəˌhəːts]	n. 兆赫(MHz)	10A
mesh	[meʃ]	n. 网孔	
meter	[ˈmiːtə]	n. 仪表,米,计,表	3A
method	[ˈmeθəm]	n. 方法	3B

续表

单　词	音　标	意　义	课次
metric	['metrik]	adj. 米制的,公制的	9B
mica	['maikə]	n. 云母	1A
microprocessor	[maikrəu'prəusesə]	n. 微处理器	8A
microscopic	[,maikrə'skɔpik]	adj. 用显微镜可见的,精微的	2A
microvolt	[,maikrəu'vəult]	n. 微伏[等于1伏(特)的百万分之一]	2A
migrate	[mai'greit, 'maigreit]	vi. 迁移,移动,移往,移植 vt. 使移居,使移植	4B
millimeter	['milimi:tə]	n. 毫米(mm)	9B
millivolt	['milivəult]	n. 毫伏(特)[=1/1000伏(特),略作mV]	2A
miscellaneous	[misi'leinjəs, -niəs]	adj. 各色各样混在一起	9B
mismatch	['mis'mætʃ]	n. 失配	
mixer	['miksə]	n. 搅拌器,搅拌机;混频器	6A
mobility	[məu'biliti]	n. 移动性,流动性	10B
mock-up	['mɔkʌp]	n. 实验或教学用的实物大模型	8B
modem	['məudəm]	n. 调制解调器	10B
moderate	['mɔdərit]	adj. 中等的,适度的,适中的;vt. & vi. 缓和	4B
modulator	['mɔdjuleitə]	n. 调节器,调制器	
module	['mɔdju:l]	n. 模块	8A
momentary	['məuməntəri]	adj. 瞬间的,刹那间的	5B
monitor	['mɔnitə]	n. 监听器,监视器,监控器;vt. & vi. 监控	8A
Monte Carlo	['mɔnti 'kɑ:ləu]	n. 蒙特卡洛	8B
motor	['məutə]	n. 发动机,电动机	
motoring	['məutəriŋ]	n. 电动机驱动	
multidrop	['mʌltidrɔp]	n. 多站,多支路	10B
multiple	['mʌltipl]	n. 倍数,若干	3B
multiplexer	['mʌlti,pleksə]	n. 多路(复用)器	5B
multiply	['mʌltiplai]	vt. & vi. 乘,增加	5A
narrow	['nærəu]	n. 狭窄部分,隘路 adj. 狭窄的	2A
negative	['negətiv]	n. 否定,负数; adj. 否定的,消极的,负的,阴性的; vt. 否定,拒绝(接受)	3A
netlist	['netlist]	n. 连线表,网络表	8B
network	['netwə:k]	n. 网络	
nichrome	['naikrum]	n. 镍铬合金	1A
nickel	['nikl]	n. 镍	1A
node	[nəud]	n. 节点,结点	10B
noload	['nəuləud]	adj. 空载的	
nonrotational	['nɔnrəu'teiʃənl]	adj. 无旋的	6B
nucleus	['nju:kliəs]	n. 核子	2A
numerator	['nju:məreitə]	n. 分子	3B
numerical	[nju(:)'merikəl]	adj. 数字的,用数表示的	5A
object	['ɔbdʒikt]	n. 物体,物品,实物;对象	6A
octal	['ɔktl]	adj. 八进制的	

续表

单　　词	音　　标	意　　义	课次
offset	[ˈɔːfset]	n. 偏移量	2B
ohm	[əum]	n. 欧姆	2A
ohmic	[ˈəumik]	adj. 欧姆的	2A
ohmmeter	[ˈəum,miːtə]	n. 欧姆计,电阻表	2A
oppose	[əˈpəuz]	v. 抵制,反对	2A
opposite	[ˈɔpəzit]	adj. 相反的,相对的,对面的,对立的	6A
optical	[ˈɔptikəl]	adj. 光学的	10A
orbit	[ˈɔːbit]	n. 轨道;轨迹	2A
orientation	[ˌɔ(ː)rienˈteiʃən]	n. 方向,方位,定位,倾向性,向东方	5B
oscillation	[ˌɔsiˈleiʃən]	n. 振荡,振动	1B
oscillator	[ˈɔsileitə]	n. 振荡器	1B
oscilloscope	[ɔˈsiləskəup]	n. 示波器	
outline	[ˈautlain]	n. 大纲,轮廓,略图,外形,要点,概要 vt. 描画轮廓,略述	7B
oven	[ˈʌvn]	n. 烤箱	1A
overcome	[ˌəuvəˈkʌm]	vt. 克服	6A
overhauling	[ˌəuvəˈhɔːliŋ]	n. 大修,检修	
overlapping	[ˌəuvəˈlæpiŋ]	vt. & vi. 重叠绕包,搭接	2B
overly	[ˈəuvəli]	adv. 过度地,太,非常	9A
override	[ˌəuvəˈraid]	vt. 不顾,无视,蔑视	8A
oversimplify	[ˌəuvəˈsimplifai]	vt. & vi. (使)过分地单纯化	4A
overvoltage	[ˌəuvəˈvəultidʒ]	n. 超电压,过电压	
packaging	[ˈpækidʒiŋ]	n. 包装	8A
parasitic	[ˌpærəˈsitik]	adj. 寄生的	8B
passband	[ˈpɑːsbænd]	n. 通频带,传输频带	
peak	[piːk]	n. 顶点	2B
phase	[feiz]	n. 相,相位	2B
phasor	[ˈfeizə]	n. 相量	
phenomenon	[fiˈnɔminən]	n. 现象	10A
phosphorus	[ˈfɔsfərəs]	n. 磷	4A
photodiode	[ˌfəutəuˈdaiəud]	n. 光敏二极管,光电二极管	
photolithographic	[ˌfəutə,liθəˈgræfik]	adj. 照相平版印刷(法)的	8B
photon	[ˈfəutɔn]	n. 光子	
photosensitive	[ˌfəutəuˈsensitiv]	adj. 感光性的	10A
pin	[pin]	n. 钉,销,栓,大头针,别针,腿 vt. 钉住,别住,阻止,扣牢,止住,牵制	5B
pivot	[ˈpivət]	n. 枢轴,支点	6A
plot	[plɔt]	vt. 绘图	8B
plugging	[ˈplʌgiŋ]	v. 反向制动	
pneumatic	[njuː(ː)ˈmætik]	adj. 装满空气的,有气胎的,风力的	8B
pneumatic	[njuː(ː)ˈmætik]	n. 气动	
polarity	[pəuˈlæriti]	n. 极性	2B

单词	音标	意义	课次
pole	[pəul]	n. 极,磁极,电极,极点	2B
polyphase	['pɔlifeiz]	adj. 多相的	
positive	['pɔzətiv]	adj. 阳的；adj. 肯定的,积极的,绝对的,确实的；adj. [数]正的	3A
potential	[pə'tenʃ(ə)l]	adj. 势的,位的	2A
potential	[pə'tenʃəl]	n. 电位	
power	['pauə]	n. [数]幂	5A
precise	[pri'sais]	adj. 精确的,准确的；n. 精确	9B
precision	[pri'siʒən]	n. 精确,精密度,精度	2A
precursor	[pri(:)'kə:sə]	n. 先驱	8B
predictable	[pri'diktəb(ə)l]	adj. 可预言的	9B
prefix	['pri:fiks]	n. 前缀	2A
preparatory	[pri'pærətəri]	adj. 预备的	9B
prevalent	['prevələnt]	adj. 普遍的,流行的	10B
principle	['prinsəpl]	n. 法则,原则,原理	7A
product	['prɔdəkt]	n. 乘积	6A
programmable	['prəugræməbl]	adj. 可设计的,可编程的	8A
propagate	['prɔpəgeit]	v. 传导,传播	
property	['prɔpəti]	n. 性质,特性	6A
proportional	[prə'pɔːʃənl]	adj. 比例的,成比例的,相称的	2A
proportionality	[prə,pɔːʃə'næliti]	n. 比例(性)	2A
pros	[prəuz]	adv. 正面地	10A
protocol	['prəutəkɔl]	n. 协议	8A
proton	['prəutɔn]	n. 质子	2A
pulley	['puli]	n. 滑车,滑轮	9A
pump	[pʌmp]	n. 泵,抽水机；vt. (用泵)抽(水),抽吸	6A
pushbutton	[puʃ'bʌtn]	n. 按钮	8B
pyramid	['pirəmid]	n. 角锥,棱锥	4A
questionable	['kwestʃənəb(ə)l]	adj. 有问题的,不肯定的,靠不住的	10B
quiescent	[kwai'esənt]	adj. 静止的	5B
quill	[kwil]	n. 钻轴	9A
radiator	['reidieitə]	n. 散热器,冰箱	1A
radius	['reidjəs]	n. 半径,范围,界限	6A
ramp	[ræmp]	n. 斜坡,坡道；vi. 蔓延；vt. 使有斜面	5B
rate	[reit]	n. 比率,速度,等级	2A
rated	['reitid]	adj. 额定的	
ratio	['reiʃiəu]	n. 比,比率,变比	6A
reactance	[ri'æktəns]	n. 电抗	
reactive	[ri(:)'æktiv]	adj. 电抗的,无功的	
reactor	[ri(:)'æktə]	n. 电抗器	
reciprocal	[ri'siprəkəl]	adj. 倒数的 n. 倒数	2A
reconnect	[,ri:kə'nekt]	vt. & vi. 再接合	8A

续表

单词	音标	意义	课次
rectangular	[rek'tæŋgjulə]	adj. 矩形的,成直角的	9B
rectifier	['rektifaiə]	n. 整流器,矫正器	1B
redirect	[ˌriːdi'rekt]	vt. (信件)重寄,使改道,使改方向	10A
redraw	[ˌriː'drɔː]	vt. 重画;vi. 刷新(屏幕)	3B
reduce	[ri'djuːs]	vt. 减少,缩小,简化,还原	3B
regeneration	[riˌdʒenə'reiʃən]	n. 再生,后反馈放大	
register	['redʒistə]	n. 寄存器	
regulation	[ˌregju'leiʃən]	n. 调节	
regulator	['regjuleitə]	n. 调节器	
relationship	[ri'leiʃənʃip]	n. 关系,关联	3A
relatively	['relətivli]	adv. 相关地	7A
relay	['riːlei]	n. 继电器	8A
relevant	['relivənt]	adj. 有关的,相应的	7B
reluctance	[ri'lʌktəns]	n. 磁阻	
remain	[ri'mein]	vi. 剩余,残存	2A
remainder	[ri'meində]	n. 余数;adj. 剩余的	5A
repeater	[ri'piːtə]	n. 转发器,中继器	10B
repel	[ri'pel]	vt. 排斥	2A
represent	[ˌriːpri'zent]	vt. 表示,表现	2A
repulsion	[ri'pʌlʃən]	n. 排斥	2A
resemble	[ri'zembl]	vt. 像,类似	8B
reset	['riːset]	v. 复位	
residential	[ˌrezi'denʃəl]	adj. 住宅的,民用的	6A
resist	[ri'zist]	n. 阻抗	1A
resistance	[ri'zistəns]	n. 阻力,电阻,阻抗	1A
resistance	[ri'zistəns]	n. 电阻	
resistivity	[ˌriːzis'tiviti]	n. 电阻系数	2A
resistor	[ri'zistə]	n. 电阻器	
resistor	[ri'zistə]	n. 电阻器	2A
resonance	['rezənəns]	n. 谐振,共振	
resonate	['rezəneit]	n. 共振,共鸣;vt. 谐振,共鸣,回响,调谐	1B
response	[ris'pɔns]	n. 回答,响应,反应	4B
resultant	[ri'zʌltənt]	adj. 作为结果而发生的,合成的	6B
rever	[ri'viə, eə'vɛə]	n. 转子	2B
reverse	[ri'vəːs]	n. 相反,背面,反面,倒退;adj. 相反的,倒转的,颠倒的;vt. 颠倒,倒转	4B
revolution	[ˌrevə'luːʃən]	n. 旋转	6A
rheostat	['riːəˌstæt]	n. 变阻器,电位器	
robust	[rə'bʌst]	adj. 健壮的	8B
rotatable	['rəuteitəbl]	adj. 可旋转的,可转动的	7A
rotate	[rəu'teit]	vt. & vi. (使)旋转	6A

续表

单　　词	音　　标	意　　义	课次
rotor	['rəutə]	n. 转子	
rubber	['rʌbə]	n. 橡皮,橡胶	2A
sample	['sæmpl]	n. 标本,样品,例子; vt. 取样,采样,抽取……的样品	7B
saturation	[,sætʃə'reiʃn]	n. 饱和	
scale	[skeil]	n. 刻度,量程	
schematic	[ski'mætik]	adj. 示意性的 n. 电路原理图	3A
scheme	[ski:m]	n. 安排,配置,计划,方案,图解,摘要; vt. & vi. 计划,设计,图谋,策划	5B
screw	[skru:]	n. 螺钉,螺旋,螺杆,螺孔; vt. 调节,旋,加强,压榨,强迫,鼓舞; vi. 转动,旋,拧	9B
semiconductor	['semikən'dʌktə]	n. 半导体	4A
sensor	['sensə]	n. 传感器	7A
series	['siəri:z]	n. 连续,串联,串励	1A
servomechanism	['sə:vəu'mekənizəm]	n. 伺服系统	
set	[set]	v. 置位	
shaft	[ʃɑ:ft]	n. 转轴	
shape	[ʃeip]	n. 外形,形状	2A
shielding	['ʃi:ldiŋ]	adj. 防护的,屏蔽的	
shortage	['ʃɔ:tidʒ]	n. 不足,缺乏	2A
shunt	[ʃʌnt]	n. 分路器,分流器; v. 并励	
Siemens	['si:mənz]	西门子(德国电气公司)	8A
siemens	['si:mənz]	n. 西门子(欧姆的倒数)	2A
sigma	['sigmə]	n. 西格玛,希腊字母($\sum$,σ)	2A
signal	['signl]	n. 信号; adj. 信号的; vt. & vi. 发信号,用信号通知	7B
silica	['silikə]	n. 硅石,二氧化硅	
silicon	['silikən]	n. 硅,硅元素	4A
simplify	['simplifai]	vt. 单一化,简单化	3B
simulate	['simjuleit]	vt. 模拟,模仿	7B
simulator	['simjuleitə]	n. 模拟器	8B
sine	[sain]	n. 正弦	2B
slip	[slip]	n. 转差率	
slope	[sləup]	n. 斜率	
slope	[sləup]	n. 斜坡,斜面,倾斜	2A
slot	[slɔt]	n. 槽	
solenoid	['səulinɔid]	n. 螺线管	8B
sophisticated	[sə'fistikeitid]	adj. 富有经验的;老练的,练达;高度发展的,精密复杂的	9B
source	[sɔ:s]	n. 源极	

续表

单　词	音　标	意　义	课次
sparkover	[spɑːkˈəuvə]	v. 放电	
spectrum	[ˈspektrəm]	n. 光谱,频谱	10B
spindle	[ˈspindl]	n. 主轴,心轴,轴,杆	9A
spotty	[ˈspɔti]	adj. 质量不一的	10B
spring	[spriŋ]	n. 弹簧,发条；弹性,弹力	7A
stability	[stəˈbiliti]	n. 稳定性	
stabilizer	[ˈsteibilaizə]	n. 稳定器	
standardization	[ˌstændədaiˈzeiʃən]	n. 标准化	8B
starter	[ˈstɑːtə]	n. 启动器,启动钮	8A
statement	[ˈsteitmənt]	n. 语句	8B
stator	[ˈsteitə]	n. 定子,固定片	6B
stator	[ˈsteitə]	n. 定子	
stem	[stem]	n. 茎,干	1A
stray	[strei]	n. 杂散电容(偶然出现的间层)	1B
structure	[ˈstrʌktʃə]	n. 结构,构造	7B
style	[stail]	n. 类型,式样,字体	9A
substance	[ˈsʌbstəns]	n. 物质	2A
substation	[ˈsʌbsteiʃən]	n. 变电站	
succeeding	[səkˈsiːdiŋ]	adj. 以后的,随后的	5A
successor	[səkˈsesə]	n. 继承者,接任者	8B
sufficient	[səˈfiʃənt]	adj. 充分的,足够的	7A
suitable	[ˈsjuːtəbl]	adj. 适当的,相配的	7B
summarize	[ˈsʌməraiz]	vt. & vi. 概述,总结,摘要	10B
superconductor	[ˌsjuːpəkənˈdʌktə]	n. 超导(电)体	2A
suppression	[səˈpreʃən]	n. 抑制	10A
surge	[səːdʒ]	n. 巨涌,汹涌,澎湃；vi. 汹涌,澎湃,振荡,滑脱,放松；vt. 使汹涌奔腾,急放	10A
surge	[səːdʒ]	n. 冲击,过电压	
sweeping	[ˈswiːpiŋ]	n. 扫描	8B
swing	[swiŋ]	vt. & vi. 摇摆,摆动,回转,旋转 n. 摇摆,摆动	10B
switch	[switʃ]	n. 开关,电闸,转换	3A
switchboard	[ˈswitʃbɔːd]	n. 配电盘,开关屏	
symbol	[ˈsimbəl]	n. 符号,记号	2A
symbolically	[simˈbɔlikli]	adv. 象征性地	2A
synchronization	[ˌsiŋkrənaiˈzeiʃən]	n. 同步	
tank	[tæŋk]	n. 箱体	
tap	[tæp]	n. 分接头	
tedious	[ˈtiːdiəs]	adj. 单调乏味的,沉闷的,冗长乏味的	9A
Teletype	[ˈtelitaip]	n. 电传打字机(商标名称)	10B
temperature	[ˈtempritʃə]	n. 温度	2A
terminal	[ˈtəːminl]	n. 终端,接线端,电路接头	2A

续表

单　词	音　标	意　义	课次
terminate	[ˈtəːmineit]	vt. & vi. 停止,结束,终止	6B
terminology	[ˌtəːmiˈnɔlədʒi]	n. 术语学	6A
tetrahedron	[ˌtetrəˈhedrən]	n. 四面体	4A
theorem	[ˈθiərəm]	n. 定理,法则	5B
thermistor	[θəːˈmistə]	n. 热敏电阻	
thermostat	[ˈθəːməstæt]	n. 自动调温器,温度调节装置	7A
thick	[θik]	adj. 厚的,粗的	2A
thin	[θin]	adj. 薄的,细的	2A
threshold	[ˈθreʃəuld]	n. 临界值,阈值,门限值,门槛	10B
thrust	[θrʌst]	n. 插	9B
thumb	[θʌm]	n. 拇指	6B
thyristor	[θaiˈristə]	n. 半导体闸流管	
timer	[ˈtaimə]	n. 定时器	
timing	[ˈtaimiŋ]	n. 计时,定时器	8A
toaster	[ˈtəustə]	n. 烤炉	1A
torque	[tɔːk]	n. 扭矩,转矩,力矩	6A
touch	[tʌtʃ]	n. 触,触觉,接触; vt. 接触,触摸,触及,达到; vi. 触摸,接近,涉及,提到	9A
transceiver	[trænˈsiːvə]	n. 无线电收发机,收发器	10B
transcendental	[ˌtrænsenˈdentl]	adj. 先验的,超越的,超出人类经验的	7B
transconductance	[ˌtrænskənˈdʌktəns]	n. 跨导	
transform	[trænsˈfɔːm]	vt. 转换,改变,改造,使……变形 vi. 改变,转化,变换; n. 变换(式)	7B
transformer	[trænsˈfɔːmə]	n. 变压器	
transient	[ˈtrænziənt]	adj. 短暂的,瞬时的; n. 瞬时现象	10A
transistor	[trænˈsistə]	n. 晶体管	4A
translation	[trænsˈleiʃən]	n. 转化,转换,平移,翻译	5B
transmitter	[trænzˈmitə]	n. 传感器,传送者,变送器,发送器,传递器	10A
travel	[ˈtrævl]	n. 行程,冲程,动程	6A
triangle	[ˈtraiæŋgl]	n. 三角形	3A
trigonometric	[ˌtrigənəˈmetrik]	adj. 三角法的,据三角法的	2B
trimmer	[ˈtrimə]	n. 调整片,微调电容器	1B
tuner	[ˈtjuːnə]	n. 调谐器,调谐电路	
turn	[təːn]	n. 匝	
tutorial	[tjuːˈtɔːriəl]	n. 指南	1B
twist	[twist]	n. 一扭,扭曲,螺旋状; vt. 拧,扭曲,绞,搓,编织; vi. 扭弯,扭曲,缠绕,扭动	10A
twisting	[ˈtwistiŋ]	n. 扭曲	6A
unimaginably	[ˌʌniˈmædʒinəbl]	adj. 不能想象的,难以理解的	2A
unique	[juːˈniːk]	adj. 唯一的,独特的	8B
usage	[ˈjuːzidʒ]	n. 使用,用法	9B
utility	[juːˈtiliti]	n. 公用事业	6A

续表

单　　词	音　　标	意　　义	课次
utilize	[ˈjuːtilaiz]	vt. 利用	2A
vacuum	[ˈvækjuəm]	n. 真空,空间,真空吸尘器； adj. 真空的,产生真空的,利用真空的	4B
valid	[ˈvælid]	adj. 有效的,有根据的,正确的	7B
valve	[vælv]	n. 阀	8B
vary	[ˈvɛəri]	vt. 改变,变更,使多样化； vi. 变化,不同,违反	3A
vector	[ˈvektə]	n. 向量,矢量	6A
vector	[ˈvektə]	n. 向量, 矢量	8B
velocity	[viˈlɔsiti]	n. 速度,速率,迅速,周转率	7B
verify	[ˈverifai]	vt. 检验,校验,查证,核实	7B
verify	[ˈverifai]	vt. 检验,校验,查证,核实	2A
versus	[ˈvəːsəs]	prep. 对……；与……相对	9B
vertical	[ˈvəːtikəl]	adj. 垂直的；n. 垂直线,垂直面,竖向	2B
vertical	[ˈvəːtikəl]	adj. 垂直的,直立的,顶点的 n. 垂直线,垂直面,竖向	8B
viable	[ˈvaiəbl]	adj. 可行的,实际的,可实施的	10B
vise	[vais]	n. 老虎钳	9B
visualize	[ˈvizjuəlaiz]	vt. 形象,形象化,想象；vi. 显现	10A
Volt	[vəult, vɔlt]	n. 伏特	
voltage	[ˈvəultidʒ]	n. 电压,伏特数	2A
voltmeter	[ˈvəultˌmiːtə]	n. 电压表	3A
Watt	[wɔt]	n. 瓦特	
wattage	[ˈwɔtidʒ]	n. 瓦特数	1A
waveform	[ˈweivfɔːm]	n. 波形	2B
wavelength	[ˈweivleŋθ]	n. 波长	1B
weight	[weit]	n. 权重,位权,重量,分量	5A
winder	[ˈwaində]	n. 卷扬机	6A
winding	[ˈwaindiŋ]	n. 绕,缠,绕组,线圈	6B
wireless	[ˈwaiəlis]	adj. 无线的	10B
withstand	[wiðˈstænd]	vt. 经受住,抵挡	2B
word	[wəːd]	n. 字,词	5A
work	[wəːk]	n. 功	6A
workpiece	[ˈwəːkpiːs]	n. 工件,加工件	9A
workstation	[ˈwəːksteiʃ(ə)n]	n. 工作站	8B

词组表

词 组	意 义	课次
3-dimensional	三维的	4A
a variety of	多种的	2A
AC motor	交流马达	6A
AC transmission system	交流输电系统	
active filter	有源滤波器	
active load	有功负载	
active loss	无功损耗	
active power	有功功率	
active two-terminal network	有源二端网络	
adding counter	加法计数器	
adhere to	坚持	7A
adjoint matrix	伴随矩阵	8B
air-gap flux	气隙磁通	
air-gap line	气隙磁化线	
air-operated	气控	8B
alternating current	交流电	
alternating current component	交流分量	
alternating current path	交流通路	
amplifier region	放大区	
amplitude modulation (AM)	调幅	
amplitude-frequency response characteristic	幅频特性	
analog electronic circuit	模拟电子电路	8B
analog signal	模拟信号	8A
AND gate	与门	
angle stability	功角稳定	
angular frequency	角频率	
angular speed	角速度	6A
antilog amplifier	反对数放大器	
apparent power	视在功率	
apply to	将……应用于	3B
arc discharge	电弧放电	
arc suppression coil	消弧线圈	
armature	电枢	
armature circuit	电枢电路	
armature field	电枢场	6B
associate with	与……相关,同……联系在一起	6A
astable oscillator	非稳态振荡器	
asynchronous counter	异步计数器	
asynchronous machine	异步电机	
attachment coefficient	附着系数	
attenuation factor	衰减系数	

续表

词　　组	意　　义	课次
automatic control	自动控制	
automatic control theory	自动控制理论	
automatic meter reading	自动抄表	
automatic voltage regulator（AVR）	自动电压调整器	
autotransformer	自耦变压器	
backup relaying	后备继电保护	
backward bias	反向偏置	
band-pass filter	带通滤波器	
band-stop filter	带阻滤波器	
bare conductor	裸导线	
Basic fundamentals of power electronics	电力电子基础	
basis of analogue electronic technique	模拟电子技术基础	
be defined as	被定义为	2A
be formed by	由……组成	1B
be referable to	可归因于，与……有关	3A
belt pulley	带轮	9A
Bessel filter	贝塞尔滤波器	
between…and	在……之间	1A
bidirectional shift register	双向移位寄存器	
binary coded decimal（BCD）	二-十进制	
binary number	二进制数	
binary system	二进制系统	5A
bipolar transistor	双极(性)晶体管，场效应晶体管	8B
bistable flip-flop	双稳态触发器	
block diagram	框图	
board-level design	板级设计	8B
Bode diagram	波特图	
Boolean algebra	布尔代数	
boost-buck	升压去磁	
branch current analysis	支路电流法	
breakaway force	起步阻力	
breakdown torque	极限转矩	
Brusless DC motor	无刷直流电机	
bubble breakdown	气泡击穿	
bus bar	母线	
bus tie breaker	母联断路器	
bushing tap grounding wire	套管末屏接地线	
Butterworth filter	巴特沃斯滤波器	
by comparison	比较	9A
bypass capacitor	旁路电容	
capacitive reactance	容抗	
capacitor bank	电容器组	

续表

词　　组	意　　义	课次
capacitor filter	电容滤波器	
carbon brush	碳刷	6B
carbon family	碳族	4A
cascade transformer	串级变压器	
cast-aluminum rotor	铸铝转子	
cathode ray oscilloscope	阴极射线示波器	
center frequency	中心频率	
charging resistor	充电电阻	
Chebyshev filter	切比雪夫滤波器	
chopper circuit	斩波电路	
circuit breaker（CB）	断路器	
circuit components	电路元件	
circuit element	电路元件	
circuit element	电路元件	8B
circuit model	电路模型	
circuit parameters	电路参数	
clock pulse	时钟脉冲	
close in	包围，封闭，渐短	1A
closed-loop	闭环	7A
closed-loop gain	闭环增益	
coaxial cable	同轴电缆	
coil winding	线圈绕组	
combination logic circuit	组合逻辑电路	
command line	命令行	8B
commercial power frequency	市电频率	2A
common-drain amplifier	共漏极放大器	
common-emitter amplifier	共射极放大器	
common-mode gain	共模增益	
common-mode rejection ratio（CMRR）	共模抑制比	
common-mode signal	共模信号	
common-source amplifier	共源极放大器	
compact model	精简模型,简化模型	8B
compare with …	与……比较	7A
composite insulation	组合绝缘	
composite insulator	合成绝缘子	
compound circuits	复合电路	3B
concern with	使关心	2A
consist in	存在于	7A
consist of	由……组成	3A
controlled rectifier	可控整流	
controlling angle	控制角	
conversion of number	数字转换	5A

续表

词 组	意 义	课次
convert…to…	把……转换为……	7B
copper loss	铜损	
copper wire	铜线	2A
counter emf	反电动势	
coupled capacitor	耦合电容	
creep distance	爬电距离,漏电距离	
critical breakdown voltage	临界击穿电压	
critical clearing time	极限切除时间	
cross-over distortion	交越失真	
cross-section	截面,断面	2A
cross-sectional area	断面面积,横截面积	2A
current amplification coefficient	电流放大系数	
current density	电流密度	2A
current flow	电流	2A
current source	电流源	
current transformer (CT)	电流互感器	
cut-off distortion	截止失真	
cut-off frequency	截止频率	
cut-off region	截止区	
damping resistor	阻尼电阻	
data comparison	数据比较	8A
DC generator	直流发电机	
DC motor	直流马达	6B
dc machine	直流电机	
dead tank oil circuit breaker	多油断路器	
dead time	死区	7A
decimal system	十进制系统	5A
degree of compensation	补偿度	
demagnetization	退磁,去磁	
demodulator	解调器	
depend on	依靠,依赖	8A
detection impedance	检测阻抗	
dielectric constant	介质常数	
dielectric loss	介质损耗	
difference frequency	差频	
differential algebraic equation	微分代数方程	8B
differential amplifier	差动放大器	
differential resistance	微分电阻,内阻	2A
differentiating circuit	微分电路	
different-mode signal	差模信号	
digital electrical technique	数字电子技术	
digital signal	数字信号	8A

续表

词　　组	意　　义	课次
digital signal processing	数字信号处理	
direct axis transient time constant	直轴瞬变时间常数	
direct current	直流电	
direct current circuit	直流电路	
direct current component	直流分量	
direct current path	直流通路	
direct-coupled amplifier	直接耦合放大器	
direct-reset terminal	直接复位端	
direct-set terminal	直接置位端	
discrete input	开关量输入	8B
discrete output	开关量输出	8B
distribution automation system	配电网自动化系统	
distribution dispatch center	配电调度中心	
distribution system	配电系统	
divide into	分为	2B
divider ratio	分压器分压比	
domestic load	民用电	
doped semiconductor	掺杂半导体	
double-column transformer	双绕组变压器	
down transformer	降压变压器	
drill press	钻床	9A
driving force	驱动力	2A
droop characteristics	下降特性	
droop rate	下降率	
drop point	落点	
dual bus	双总线	
due to	由于,应归于	9A
dynamic resistance	动态电阻	
dynamic response	动态响应	
dynamic state	动态	
earthing switch	接地开关	
eddy current	涡流	
effective value	有效值	
electric current	电流	2A
electric energy	电能	
electric field	电场	2A
electrical conductivity	电导率	2A
electrical device	电气设备	
electrical drive and control	电力传动与控制	
electrical energy	机械能	6A
electrical machinery	电机学	
electrical substation	变压站	2A

续表

词　组	意　义	课次
electrochemical deterioration	电化学腐蚀	
electromagnetic field	电磁场	
electromotive force	电动势	
electromotive force	电动势	2A
electron avalanche	电子雪崩	
electron configuration	电子构型,电子组态	2A
electronegative gas	电负性气体	
electronic machine system	电子电机集成系统	
electrostatic voltmeter	静电电压表	
emitter-follower	射极跟随器	
end ring	端环	
energy converter	电能转换器	
enhancement mode MOSFET (E-MOSFET)	增强型 MOS 场效应管	
epoxy resin	环氧树脂	
equal value resistors	等值电阻	3B
equivalent circuit	等效电路	
error detector	误差检测器	
error signal	误差信号	
excitation system	励磁系统	
exciting voltage	励磁电压	
exciting winding	激磁绕组	
expulsion gap	灭弧间隙	
external force	外力	6A
extra-high voltage (EHV)	超高压	
fall in	属于	4A
fault clearing time	故障切除时间	
feedback loop	反馈回路,反馈环	8B
feedback loop	反馈回路	
feedback path	反馈通道	
feedback signal	反馈信号	
feedback system	反馈系统	
Fiber Optic	光纤	10B
field current	励磁电流	
field distortion	场畸变	
field emission	场致发射	
field gradient	场梯度	
field strength	场强	
field stress	电场力	
field-effect transistor (FET)	场效应管	
fill with	使充满	2A
first-order circuit	一阶电路	
first-order filter	一阶滤波器	

续表

词　　组	意　　义	课次
fixed contact	静触头	
fixed series capacitor compensation	固定串联电容补偿	
flash counter	雷电计数器	
flow control	液控	8B
flow through	流过	3B
flux density	磁感应强度	
flux linkage	磁链	
force out	挤(出去),冲(出去)	2A
forward bias	正向偏置	
forward transfer function	正向传递函数	
fossil-fired power plant	火电厂	
Fourier series	傅立叶级数	
four-terminal sensing	四端子检测,四线检测,四点探针法	2A
free electron	自由电子	2A
frequency characteristic	频率特性	
frequency response	频率响应	
frequency domain	频域	
front resistance	波头电阻	
full load	满载	
full adder	全加器	
full-load torque	满载转矩	
full-wave rectifier	半波整流	
gaseous insulation	气体绝缘	
generator terminal	机端	
generator voltage	发电机电压	
germanium diode	锗二极管	
glass insulator	玻璃绝缘子	
glow discharge	辉光放电	
grading ring	均压环	
graphic representation	图示	2B
graphical user interface	图形用户界面	8B
grounding capacitance	对地电容	
half-wave rectifier	全波整流	
Hard-Wired Control	硬接线控制	8A
harmonic balance	谐波平衡	8B
heating appliance	电热器	
high limited value	高顶值	
high voltage	高压	
high voltage engineering	高电压工程	
high voltage shunt reactor	高抗	
high-pass filter	高通滤波器	
high-performance	高性能的	

续表

词　　组	意　　义	课次
highvoltage engineering	高电压工程	
highvoltage testing technology	高电压实验技术	
horseshoe magnet	马蹄形磁铁	
hydraulic turbine	水轮机	
hydro generator	水轮发电机	
hydro power station	水力发电站	
hysteresis comparator	迟滞比较器	
I/O point	输入/输出点	8A
ideal current source	理想电流源	
ideal voltage source	理想电压源	
implicit integration methods	隐式积分方法	8B
impulse current	冲击电流	
impulse flashover	冲击闪络	
in simple term	简言之	6A
in spite of	不管	4A
incoming line	进线	
independent of	不依赖……,独立于……	7A
independent pole operation	分相操作	
index finger	食指	6B
induced current	感生电流	
induction motor	感应电动机	
inductive reactance	感抗	
inductor filter	电感滤波器	
industrial bus	工业总线	10B
industrial environment	工业环境	10B
inhomogenous field	不均匀场	
initial phase	初相位	
initialization parameter	最初参数,初始参数	8B
input impedance	输入阻抗	
input resistance	输入电阻	
inrush current	涌流	
installed capacity	装机容量	
instantaneous value	瞬时值	
instrument transducer	测量互感器	
instrumentation amplifier	仪表放大器	
insulation coordination	绝缘配合	
integrated circuit	集成电路	8B
integrating circuit	积分电路	
interfere with	妨碍,干涉,干扰	2A
intermediate relay	中间继电器	
internal combustion engine	内燃机	
internal discharge	内部放电	

续表

词　　组	意　　义	课次
intrinsic semiconductor	本征半导体	
Introduction to electrical engineering	电气工程概论	
inverter station	换流站	
inverter switch	换向开关	5B
iron core	铁芯	
iron loss	铁损	
isolation amplifier	隔离放大器	
junction diode	面结型二极管	8B
kinetic energy	动能	
Kirchhoff's current law(KCL)	基尔霍夫电流定律	
Kirchhoff's voltage law(KVL)	基尔霍夫电压定律	
ladder diagram	梯形图	
Ladder Logic Diagram	逻辑梯形图	8B
laminated core	叠片铁芯	
law of inertia	惯性定律	6A
LC oscillator	LC振荡器	
leakage flux	漏磁通	
leakage reactance	漏磁电抗	
left-hand rule	左手定则	
lightning arrester	避雷器	
lightning overvoltage	雷电过电压	
lightning stroke	雷电波	
limit switch	限位开关	8B
line current	线电流	
line printer	行式打印机	8B
line trap	线路限波器	
line voltage	线电压	
linear small-signal frequency domain analysis	线性小信号频域分析	8B
linear zone	线性区	
live tank oil circuit breaker	少油断路器	
load current	负载电流	
load resistance	负载电阻	
load shedding	甩负荷	
load-saturation curve	负载饱和曲线	
locked rotor	锁定转子	
locked-rotor torque	锁定转子转矩	
log amplifier	对数放大器	
logic expression	逻辑表达式	
logic function	逻辑函数	
loop gain	环路增益	8B
loop system	环网系统	
loss angle	(介质)损耗角	

续表

词　　组	意　　义	课次
loss of synchronization	失去同步	
low cost	低成本	10B
low of switch	换路定理	
low voltage	低压	
lower cut-off frequency	下限截止频率	
lower limit	下限	
lowest common multiple	最小公倍数	3B
low-pass filter	低通滤波器	
magnetic amplifier	磁放大器	
magnetic circuit	磁路	
magnetic field	磁场	2B
magnetic lines	磁力线	6B
magnetizing current	励磁电流	
magnetizing reactance	磁化电抗	
magnetomotive force	磁通势	
main and transfer busbar	单母线带旁路	
make of	用……制造,形成,构成	2A
manual control	手动控制	
manufacturing tolerance	制造公差	8B
master-slave flip-flop	主从型触发器	
mechanical characteristic	机械特性	
mechanical energy	电能	6A
mechanical equipment	机械设备	6A
mesh current analysis	网孔电流法	
metal oxide arrester（MOA）	氧化锌避雷器	
metal-oxide semiconductor（MOS）	金属氧化物半导体	
meters per second	米每秒	6A
middle voltage	中压	
mixed divider	（阻容）混合分压器	
moment of inertia	惯性（力）矩,转动惯量	6A
monostable flip-flop	单稳态触发器	
moving blade	可动叶片	
moving contact	动触头	
multistage amplifier	多级放大器	
mutual-inductor	互感	
N region	N区	4B
nameplate	铭牌	
NAND gate	与非门	
n-channel	N沟道	
negative charge	负电荷	2A
negative direction	负向	2B
negative edge	下降沿	

续表

词 组	意 义	课次
negative feedback	负反馈	
negative ions	负离子	
negative sequence impedance	负序阻抗	
net force	净力	6A
neutral line	中线	
neutral point	中性点	
neutral state	中性状态	2A
Newton's method	牛顿法	8B
node voltage analysis	结点电位法	
noise analysis	噪声分析	8B
no-load current	空载电流	
no-load loss	空载损耗	
non-destructive testing	非破坏性试验	
nonlinear quiescent point calculation	非线性静态点计算	8B
non-uniform field	不均匀场	
NOR gate	或非门	
normally closed contact	常闭(动断)触点	
normally open contact	常开(动合)触点	
north pole	北极	6B
Norton's theorem	诺顿定理	
NOT gate	非门	
N-type	N 型	4A
N-type semiconductor	N 型半导体	
nuclear power station	核电站	
number of poles	极数	
number system	数字系统	5A
numerical value	数字值	5A
of the time	当时的,当代的	8B
offset current	失调电流	
offset voltage	失调电压	
Ohm's Law	欧姆定律	3A
Ohm's Law	欧姆定律	2A
oil-filled power cable	充油电力电缆	
oil-impregnated paper	油浸纸绝缘	
on the other hand	另一方面	1A
open circuit	开路(断路)	
open source	开放源码,开源	8B
open-circuit voltage	开路电压	
open-loop gain	开环增益	
operation amplifier	运算放大器	
operation mechanism	操动机构	
operational amplifier(op-amp)	运算放大器	

续表

词　　组	意　　义	课次
operational calculus	算符演算	
opposite direction	反向	2B
optical fiber	光纤	
OR gate	或门	
outer electron	外层电子	4A
outgoing line	出线	
output resistance	输出电阻	
over time	超时,滞后	7A
overall voltage gain	总的电压放大倍数	
overhead line	架空线	
P region	P 区	4B
parallel branch	并联分支	3B
parallel circuit	并联电路	3B
parallel negative feedback	并联负反馈	
parallel resonance	并联谐振	
parametric sweep	参数扫描	8B
partial discharge	局部放电	
passive filter	无源滤波器	
passive two-terminal network	无源二端网络	
past voltage within power system	电力系统内部过电压	
p-channel	P 沟道	
peak point	峰点	
peak reverse voltage	反向峰值电压	
peak value	峰值	2B
peak voltmeter	峰值电压表	
peak-load	峰荷	
per unit value	标幺值	
percentage	百分数	
performance characteristic	工作特性	
Periodic Table	（元素）周期表	4A
permanent magnet	永久磁铁	6B
Permanent-magnet synchronism motor	永磁同步电机	
phase current	相电流	
phase difference	相位差	
phase displacement (shift)	相移	
phase inversion	倒相,反相	
phase lag	相位滞后	
phase lead	相位超前	
phase shift	相移	
phase shifter	移相器	
phase voltage	相电压	
phase-frequency response characteristic	相频特性	

续表

词　　组	意　　义	课次
phase-locked loop (PLL)	锁相环	
phase-to-phase voltage	线电压	
phasor diagram	相量图	
photoelectric emission	光电发射	
photolithographic masks	光刻掩膜	8B
physics types	物理类型,物理结构	4A
pinch-off voltage	夹断电压	
PN junction	PN 结	4B
point plane gap	针板间隙	
polarity effect	极性效应	
pole-zero analysis	极零点分析	8B
polyphase rectifier	多相整流器	
porcelain insulator	陶瓷绝缘子	
positive charge	正电荷	2A
positive direction	正向	2B
positive edge	上升沿	
positive feedback	正反馈	
positive sequence impedance	正序阻抗	
potential difference	电位差	
potential energy	势能	
potential stress	电位应力,电场强度	
potential transformer (PT)	电压互感器	
power amplifier	功率放大器	
power angle	功率角	
power cable	电力电缆	2A
power capacitor	电力电容	
power electronics	电力电子	
power factor	功率因数	
power line carrier (PLC)	电力线载波(器)	
power network	电力网络	
power plant	电厂	
power system	电网,电力系统	
power system relaying protection	电力系统继电保护	
power transfer	能量输送	
power transformer	电力变压器	
power transmission system	输电系统	
power-factor compensation	功率因数补偿	
power-flow current	工频续流	
practical situation	实际情况	2A
pressure difference	压力差,差压	2A
pressure drop	压力降	2A
primary cell	原生电池	

续表

词　　组	意　　义	课次
primary relaying	主继电保护	
prime grid substation	主网变电站	
prime mover	原动机	
Principle of circuits	电路原理	
printed wiring board	印刷线路板	8B
principle of electrical system's relay protection	电力系统继电保护原理	
principle of microcomputer	微机原理	
principles of electric circuits	电路原理	
programmable controller	可编程控制器	5A
programmable logic controller	可编程逻辑控制器	
proportional system	比例系统	7A
proportionality constant	比例常数	2A
protection principle of power system's element	电力系统元件保护原理	
protective relaying	继电保护	
proximity effect	邻近效应	2A
proximity switch	接近开关	8B
P-type	P型	4A
P-type semiconductor	P型半导体	
pull-up resistor	牵引电阻	5B
pumped storage power station	抽水蓄能电站	
quality factor	品质因数	
quasi-uniform field	稍不均匀场	
quiescent point (Q-point)	静态工作点	
radio interference	无线干扰	
random-access memory (RAM)	随机存取存储器	
random-wound	散绕	
rated current	额定电流	
rated power	额定功率	
rated voltage	额定电压	
rating of equipment	设备额定值	
reactive current	无功电流	
reactive load	无功负载	
reactive loss	有功损耗	
reactive power	无功功率	
reactive power compensation	无功补偿	
read-only memory (ROM)	只读存储器	
recovery voltage	恢复电压	
reference direction	参考方向	
reference potential	参考电位	
reference value	参考值	
reference voltage	基准电压	
referred to as	称为	2B

续表

词　　组	意　　义	课次
regardless of	不管,不顾	2A
reinforced excitation	强行励磁	
relay panel	继电器屏	
reserve capacity	备用容量机	
residual capacitance	残余电容	
resistance-capacitance coupled amplifier	阻容耦合放大器	
resonance frequency	谐振频率	
response characteristic	频响特性(曲线)	
resulted in	导致	7B
retaining ring	护环	
reverse leakage current	反向漏电流	
revolutions per minute	转/分	
right-hand rule	右手法则	6B
right-hand rule	右手定则	
ring bus	环形母线	
Rogowski coil	罗可夫斯基线圈	
Root-Mean-Square	均方根(值)	2B
root-mean-squire value (rms)	均方根值	
rotating magnetic field	旋转磁场	
rotor core	转子铁芯	
rotor resistance	转子电阻	
routing testing	常规试验	
sampling interval	采样间隔	7B
sampling period	采样周期	7B
saturated output level	饱和输出电平(电压)	
saturation curve	饱和曲线	
saturation distortion	饱和失真	
saturation effect	饱和效应	
saturation region	饱和区	
schematic capture	获得原理图	8B
schematic diagram	原理图,示意图	8B
Schering bridge	西林电桥	
second-order filter	二阶滤波器	
self exciting	自励的	
semiconductor technology	半导体工艺	8B
sensitivity analysis	灵敏度分析	8B
separately excited	他励的	
sequential logic circuit	时序逻辑电路	
series (voltage) regulator	串联型稳压电源	
series circuit	串联电路	3A
series compensation	串联补偿	
series negative feedback	串联负反馈	

续表

词　组	意　义	课次
series resistance	串联电阻	3A
series resonance	串联谐振	
shield wire	避雷线	
shift register	移位寄存器	
short circuit	短路	
short-circuit current	短路电流	
short-circuit ratio	短路比	
short-circuiting ring	短路环	
shunt capacitor	并联电容器	
shunt compensation	并联补偿	
shunt field	并励磁场	
shunt reactor	并联电抗器	
side by side	并肩的,并行的	3B
silicon carbide	碳化硅	
silicon diode	硅二极管	
silicon rubber	硅橡胶	
silicon-controlled rectifier	可控硅	
simulation analysis	仿真分析	
sine wave	正弦波	2B
single bus	单母线	
Single Side Band (SSB)	单边带	
single-phase asynchronous motor	单相异步电动机	
sinusoidal a-c circuit	正弦交流电路	
sinusoidal oscillator	正弦波振荡器	
sinusoidal voltage	正弦电压	
skin effect	趋肤效应	2A
slip ring	滑环	
solid state	固体	
source transformations	电源变换	
south pole	南极	6B
space charge	空间电荷	
sparse matrix	稀疏矩阵	8B
speed regulation	速度调节	
speed-torque curve	转速力矩特性曲线	
square metres	平方米	2A
square wave generator	方波发生器	
squirrel cage	鼠笼	
stability calculation	稳定计算,稳性计算	8B
stabilization network	稳定网络	
stabilizing transformer	稳定变压器	
stand for	代表,代替,象征	9A
star connection(Y-connection)	星形连接	

续表

词　　组	意　　义	课次
starting current	起动电流	
starting torque	起动转矩	
statement list	语句表	
static resistance	静态电阻	2A
static state	静态	
static var compensation (SVC)	静止无功补偿	
stationary blade	固定叶片	
stator winding	定子绕组	
steady state	稳态	
steady state	恒稳态,定态	8B
steady-State analysis of power system	电力系统稳态分析	
steam engine	蒸汽机	6A
steam turbine	汽轮机	
steel wire	钢丝	2A
steel-reinforced aluminum conductor	钢芯铝绞线	
step by step	一步步,逐步地	9A
step input voltage	阶跃输入电压	
step up transformer	升压变压器	
storage battery	蓄电池	
stray capacitance	杂散电容	
stray inductance	杂散电感	
streamer breakdown	流注击穿	
subtracting counter	减法计数器	
sum frequency	和频	
superposition theorem	叠加原理	
supervisory control and data acquisition (SCADA)	监控与数据采集	
surface breakdown	表面击穿	
surge impedance	波阻抗	
suspension insulator	悬式绝缘子	
sustained discharge	自持放电	
switch station	开关站	
switching (voltage) regulator	开关型稳压电源	
switching overvoltage	操作过电压	
symmetrical three-phase load	对称三相负载	
symmetrical three-phase source	对称三相电源	
synchronous condenser	同步调相机	
synchronous counter	同步计数器	
synchronous generator	同步发电机	
synchronous motor	同步电动机	
synchronous speed	同步转速	
tail resistance	波头电阻	
take advantage of	利用	1A

续表

词　　组	意　　义	课次
take…into account	考虑	1B
tap position	档位	
taped transformer	多级变压器	
temperature extremes	温度极限	8B
tertiary winding	第三绕组	
test object	被试品	
the Bode plot	波特图	
the dielectric	电介质	
the Karnaugh map	卡诺图	
thermal breakdown	热击穿	
thermal overload relay	热继电器	
thermal power station	火力发电站	
Thevenin's theorem	戴维宁定理	
third finger	中指	6B
three phase fault	三相故障	
three-column transformer	三绕组变压器	
three-factor method	三要素法	
three-phase asynchronous motor	三相异步电动机	
three-phase circuit	三相电路	
three-phase four-wire system	三相四线制	
three-phase power	三相功率	
three-phase source	三相电源	
three-phase three-wire system	三相三线制	
thumbwheel switch	指轮开关	5A
tidal current	潮流	
time constant	时间常数	
time delay	延时	
time domain	时域	
time invariant	时不变的	
time phase	时间相位	
time relay	时间继电器	
time-domain large-signal solution	时域大信号解	8B
timing diagram	时序图	
transfer curve	转换曲线	8B
transfer function	传递函数,转移函数	8B
transfer switching	倒闸操作	
transformer substation	变电站	
transient analysis	瞬态分析,暂态分析	8B
transient response	瞬态响应	
transient response	暂态过程,暂态响应	
transient stability	暂态稳定	
transient state	暂态	

续表

词　　组	意　　义	课次
transmission lines	传输线，波导线	8B
travel switch	行程开关	
triangular connection(D-connection,delta connection)	三角形连接	
triangular wave generator	三角波发生器	
trigger electrode	触发电极	
trigger pulse	触发脉冲	
trigonometric function	三角函数	2B
trip circuit	跳闸电路	
trip coil	跳闸线圈	
tri-state gate	三态门	
truth table	真值表	5B
tuned circuit	调谐电路	
turbo generator	汽轮发电机	
turn-on angle	导通角	
two-way configuration	二线制	
ultra-high voltage(UHV)	特高压	
underground cable	地下电缆	
uniform cross section	等截面	2A
uniform field	均匀场	
unijunction transistor(UJT)	单结晶体管	
uninterruptible power supply	不间断电源	
unity-gain bandwidth	单位增益带宽	
upper cut-off frequency	上限截止频率	
upper limit	上限	
vacuum circuit breaker	真空断路器	
valley point	谷点	
variable transformer	调压变压器	
Very Large Scale Integration	超大规模集成电路	5B
voltage control system	电压控制系统	
voltage difference	电压差	10B
voltage divider	分压器	
voltage drop	电压降	3A
voltage gain	电压放大倍数	
voltage grade	电压等级	
voltage source	电压源	
voltage stability	电压稳定	
voltage-follower	电压跟随器	
volt-ampere characteristic	伏安特性	
walk around…	绕……而走	6A
warm up	变暖,热身	7A
wave front	波头	
wave guide	波导,波导管	

续表

词　　组	意　　义	课次
wave tail	波尾	
wind-driven generator	风动发电机	
withstand test	耐压试验	
withstand voltage	耐受电压	
zener diode	稳压二极管,齐纳二极管	
zero drift	零点漂移	
zero sequence current	零序电流	
zero sequence impedance	零序阻抗	
zinc oxide	氧化锌	

缩写表

缩　　写	意　　义	课次
A/D（Analog to Digital）	模拟/数字转换	
AC（Alternating Current）	交流电	6A
ADC（Analog to Digital Converter）	模拟/数字转换器	
ADM（Adaptive Delta Modulation）	自适应增量调制	
ADPCM（Adaptive Differential Pulse Code Modulation）	自适应差分脉冲编码调制	
ALU（Arithmetic Logic Unit）	算术逻辑单元	
ASCII（American Standard Code For Information Interchange）	美国信息交换标准码	
ASIC（Application-Specific Integrated Circuit）	专用集成电路	
AV（Audio Visual）	声视,视听	
BCD（Binary Coded Decimal）	二进制编码的十进制	5A
BCR（Bi-directional Controlled Rectifier）	双向晶闸管	
BCR（Buffer Courtier Reset）	缓冲计数器	
BSIM（Berkeley Short-channel IGFET Model）	伯克利短沟道 IGFET 模型	8B
BZ（buzzer）	蜂鸣器,蜂音器	
C（capacitance,capacitor）	电容量,电容器	
CATV（CAble TeleVision）	电缆电视	
CCD（Charge-Coupled Device）	电荷耦合器件	
CCTV（Closed-Circuit TeleVision）	闭路电视	
CEMF（Counter Electro Motive Force）	反电动势	6B
CMOS（Complementary Metal-Oxide-Semiconductor Transistor）	互补金属氧化物半导体	5B
CMOS（Complementary MOS）	互补 MOS	
CNC（Computerized Numerical Control）	计算机数字控制	
CPLD（Complex Programmable Logic Device）	复杂可编程逻辑器件	
CPU（Central Processing Unit）	中央处理器	
CS（Control Signal）	控制信号	

续表

缩　　写	意　　义	课次
D（diode）	二极管	
D/A（Digit/Analogy）	数字/模拟	7B
DAST（Direct Analog Store Technology）	直接模拟存储技术	
DC（Direct Current）	直流电	2B
DIP（Dual In-line Package）	双列直插封装	
DP（Dial Pulse）	拨号脉冲	
DRAM（Dynamic Random Access Memory）	动态随机存储器	
DTL（Diode-Transistor Logic）	二极管晶体管逻辑	
DUT（Device Under Test）	被测器件	
DVM（Digital VoltMeter）	数字电压表	
ECL（Emitter Coupled Logic）	射极耦合逻辑	
EDA（Electronic Design Automation）	电子设计自动化	
EDI（Electronic Data Interchange）	电子数据交换	
EEPROM（Electrically EPROM）	电可擦可编程只读存储器	
EIA（Electronic Industries Association）	电子工业联合会	10B
EMC（Electromagnetic Compatibility）	电磁兼容	
EOC（End Of Conversion）	转换结束	
EPROM（Erasable Programmable Read Only Memory）	可擦可编程只读存储器	
ESD（Electro-Static Discharge）	静电放电	
F/V（Frequency to Voltage convertor）	频率/电压转换	
FAM（Pulse Amplitude Modulation）	脉冲幅度调制	
FCC（Federal Communications Commission）	（美国）通信委员会	10B
FET（Field-Effect Transistor）	场效应晶体管	
FFT（Fast Fourier Transform）	快速傅里叶变换	
FM（Frequency Modulation）	调频	
FPGA（Field Programmable Gate Array）	现场可编程门阵列	
FS（Full Scale）	满量程	
FSK（Frequency Shift Keying）	频移键控	
FSM（Field Strength Meter）	场强计	
FST（Fast Switching Shyster）	快速晶闸管	
FT（Fixed Time）	固定时间	
FU（Fuse Unit）	保险丝装置	
FWD（forward）	正向的	
GAL（Generic Array Logic）	通用阵列逻辑	
GIS（Gas Insulated Substation）	气体绝缘变电站	
GND（ground）	接地,地线	
GTO（Gate Turn-Off transistor）	门极可关断晶体管	
HART（Highway Addressable Remote Transducer）	高速可寻址远程传感器	
HDL（Hardware Description Language）	硬件描述语言	
HF（High Frequency）	高频	
hp（horsepower）	马力	6A
HTL（High Threshold Logic）	高阈值逻辑电路	

续表

缩　　写	意　　义	课次
HTS（Heat Temperature Sensor）	热温度传感器	
I/O（Input/Output）	输入/输出	
I/V（Current to Voltage Convertor）	电流/电压变换器	
IC（Integrated Circuit）	集成电路	
ID（International Data）	国际数据	
IEC（International Electrotechnical Commission）	国际电工（技术）委员会	
IEE（Institution of Electrical Engineers）	电气工程师学会（英）	
IEEE（Institute of Electrical and Electronic Engineers）	电气与电子工程师学会（美）	
IGBT（Insulated Gate Bipolar Transistor）	绝缘栅双极型晶体管	
IGFET（Insulated Gate Field Effect Transistor）	绝缘栅场效应晶体管	
IP（Intelligent Property）	智能模块	
IPM（Incidental Phase Modulation）	附带的相位调制	
IPM（Intelligent Power Module）	智能功率模块	
IR（Infrared Radiation）	红外辐射	
IRQ（Interrupt ReQuest）	中断请求	
IS（International System）	国际系统	2A
ISO（International Standardization Organization）	国际标准化组织	
JFER（Junction Field Effect Transistor）	结栅场效应晶体管	8B
JFET（Junction Field Effect Transistor）	结型场效应晶体管	
LAN（Local Area Network）	局域网	
LAS（Light Activated Switch）	光敏开关	
LASCS（Light Activated Silicon Controlled Switch）	光控可控硅开关	
LCD（Liquid Crystal Display）	液晶显示器	
LDC（Line Drop Compensation）	线路补偿器	
LDR（Light Dependent Resistor）	光敏电阻	
LED（Light Emitting Diode）	发光二极管	
LRC（Longitudinal Redundancy Check）	纵向冗余（码）校验	
LSB（Least Significant Bit）	最低有效位	
LSI（Large Scale Integration）	大规模集成电路	
M（motor）	电动机	
MCT（MOS Controlled Gyrator）	场控晶闸管	
MIC（microphone）	话筒，微音器，麦克风	
min（minute）	分	
MOS（Metal Oxide Semiconductor）	金属氧化物半导体	
MOSFET（Metallic Oxide Semiconductor Field Effect Transistor）	金属氧化物半导体场效应晶体管	8B
MOSFET（Metal Oxide Semiconductor FET）	金属氧化物半导体场效应晶体管	
N（negative）	负	
NMOS（N-Channel Metal Oxide Semiconductor）	N沟道MOS	
NTC（Negative Temperature Coefficient）	负温度系数	
OC（Over Current）	过电流	
OCB（Overload Circuit Breaker）	过载断路器	

续表

缩　　写	意　　义	课次
OCS（Optical Communication System）	光通信系统	
OR（type of logic circuit）	或逻辑电路	
OV（Over Voltage）	过电压	
P（pressure）	压力	
PAL（Programmable Array Logic）	可编程阵列逻辑	
PC（Pulse Code）	脉冲码	
PCM（Pulse Code Modulation）	脉冲编码调制	
PDM（Pulse Duration Modulation）	脉冲宽度调制	
PF（Power Factor）	功率因数	
PFM（Pulse Frequency Modulation）	脉冲频率调制	
PG（Pulse Generator）	脉冲发生器	
PGM（Programmable）	编程信号	
PI（Proportional Integral）	比例积分	
PID（Proportional Integral Differential）	比例、积分、微分（控制器）	7B
PIN（Positive Intrinsic-Negative）	光电二极管	
PIO（Parallel Input Output）	并行输入输出	
PLC（Programmable Logic Controller）	可编程序逻辑控制器	8A
PLD（Phase-Locked Detector）	同相检波	
PLD（Phase-Locked Discriminator）	锁相解调器	
PLL（Phase-Locked Loop）	锁相环路	
PMOS（P-channel Metal Oxide Semiconductor）	P沟道MOS	
P-P（Peak-to-Peak）	峰-峰	
PPI（Point-to-Point Interface）	点-点接口	8A
PPM（Pulse Phase Modulation）	脉冲相位洲制	
PRD（Piezoelectric Radiation Detector）	热电辐射探测器	
PROM（Programmable Read Only Memory）	可编只读程存储器	
PRT（Platinum Resistance Thermometer）	铂电阻温度计	
PRT（Pulse Recurrent Time）	脉冲周期时间	
PUT（Programmable Unijunction Transistor）	可编程单结晶体管	
PWM（Pulse Width Modulation）	脉宽调制	
R（resistance,resistor）	电阻,电阻器	
RAM（Random Access Memory）	随机存储器	
RCT（Reverse Conducting Thyristor）	逆导晶闸管	
REF（reference）	参考,基准	
REV（reverse）	反转	
RF（Radio Frequency）	射频,无线电频率	10B
RGB（Red/Green/Blue）	红绿蓝	
ROM（Read Only Memory）	只读存储器	
RP（Resistance Potentiometer）	电位器	
RPM（Revolutions Per Minute）	转/分	6A
RST（reset）	复位信号	
RT（resistor with inherent variability dependent）	热敏电阻	

续表

缩　　写	意　　义	课次
RTD（Resistance Temperature Detector）	电阻温度传感器	
RTL（Register Transfer Level）	寄存器传输级描述	
RTL（Resistor Transistor Logic）	电阻晶体管逻辑（电路）	
SA（Switching Assembly）	开关组件	
SBS（Silicon Bi-directional Switch）	硅双向开关，双向硅开关	
SCR（Silicon Controlled Rectifier）	可控硅整流器	
SCS（Safety Control Switch）	安全控制开关	
SCS（Silicon Controlled Switch）	可控硅开关	
SCS（Speed Control System）	速度控制系统	
SCS（Supply Control System）	电源控制系统	
SG（Spark Gap）	放电器	
SIT（Static Induction Transformer）	静电感应晶体管	
SITH（Static Induction Thyristor）	静电感应晶闸管	
SLIC（simulator for linear integrated circuits）	线性集成电路模拟程序	8B
SLIC（System Level IC）	系统级 IC	
SOC（System On a Chip）	片上系统	
SP（Shift Pulse）	移位脉冲	
SPI（Serial Peripheral Interface）	串行外围接口	
SPICE（Simulation Program with Integrated Circuit Emphasis）	集成电路模拟程序	8B
SR（Sample Relay，Saturable Reactor）	取样继电器，饱和电抗器	
SR（Silicon Rectifier）	硅整流器	
SRAM（Static Random Access Memory）	静态随机存储器	
SSR（Solid-State Relay）	固体继电器	
SSR（Switching Select Repeater）	中断器开关选择器	
SSS（Silicon Symmetrical Switch）	硅对称开关，双向可控硅	
SSW（Synchrony-Switch）	同步开关	
ST（start）	启动	
ST（starter）	启动器	
STB（strobe）	闸门，选通脉冲	
T（transistor）	晶体管，晶闸管	
TACH（tachometer）	转速计，转速表	
TP（Temperature Probe）	温度传感器	
TRIAC（TRIodes AC switch）	三极管交流开关	
TTL（Transistor Transistor Logic）	晶体管-晶体管逻辑	
UART（Universal Asynchronous Receiver Transmitter）	通用异步收发器	
UIC（Use Initial Conditions）	使用初始条件	8B
V/F（Voltage-to-Frequency）	电压/频率转换	
V/I（voltage to current converter）	电压/电流变换器	
VCO（Voltage Controlled Oscillator）	压控振荡器	
VD（Video Decoders）	视频译码器	
VDR（Voltage Dependent Resistor）	压敏电阻	
VF（Video Frequency）	视频	

续表

缩　写	意　义	课次
VHDL（Very high speed integrated circuit Hardware Description Language）	超高速集成电路硬件描述语言	
VM（voltmeter）	电压表	
VS（Vacuum Switch）	电子开关	
VT（Video Terminal）	视频终端	
VT（Visual Telephone）	电视电话	
CLPE（Cross Linked Polyethylene ）	交联聚乙烯（电缆）	

注："课次"栏的标号表明在本书中出现该词汇的课次,1A 表示"第一单元的 Text A",1B 表示"第一单元的 Text B",以此类推。没有标号的为特别收入的行业内常用词汇。